# Thermal Processing of Foods

## *Control and Automation*

# IFT PRESS

The *IFT Press* series reflects the mission of the Institute of Food Technologists – to advance the science of food contributing to healthier people everywhere. Developed in partnership with Wiley-Blackwell, *IFT Press* books serve as leading-edge handbooks for industrial application and reference and as essential texts for academic programs. Crafted through rigorous peer review and meticulous research, *IFT Press* publications represent the latest, most significant resources available to food scientists and related agriculture professionals worldwide.

Founded in 1939, the Institute of Food Technologists is a nonprofit scientific society with 22,000 individual members working in food science, food technology, and related professions in industry, academia, and government. IFT serves as a conduit for multidisciplinary science thought leadership, championing the use of sound science across the food value chain through knowledge sharing, education, and advocacy.




WILEY-BLACKWELL
A John Wiley & Sons, Ltd., Publication

# Thermal Processing of Foods

## *Control and Automation*

*EDITED BY*

K.P. Sandeep
North Carolina State University
Raleigh, NC

WILEY-BLACKWELL
A John Wiley & Sons, Ltd., Publication

IFT PRESS

Edition first published 2011


Blackwell Publishing was acquired by John Wiley & Sons in February 2007. Blackwell's publishing program has been merged with Wiley's global Scientific, Technical, and Medical business to form Wiley-Blackwell.

*Editorial Office*
2121 State Avenue, Ames, Iowa 50014-8300, USA

For details of our global editorial offices, for customer services, and for information about how to apply for permission to reuse the copyright material in this book, please see our Website at www.wiley.com/wiley-blackwell.



Library of Congress Cataloging-in-Publication Data

Thermal processing of foods : control and automation / edited by K.P. Sandeep.
p. cm. – (IFT Press series)
Includes bibliographical references and index.
ISBN 978-0-8138-1007-2 (hardback)
1. Food–Preservation. 2. Food–Effect of heat on. 3. Automation. I. Sandeep, K. P.
TP371.2.T442 2011
664′.028–dc22

2010040521

A catalog record for this book is available from the U.S. Library of Congress.

Set in 11.5/13.5 Times NR PS by Aptara® Inc., New Delhi, India
Printed and bound in Singapore by Markono Print Media Pte Ltd



1 2011

## Titles in the *IFT Press* series

- *Accelerating New Food Product Design and Development* (Jacqueline H. Beckley, Elizabeth J. Topp, M. Michele Foley, J.C. Huang, and Witoon Prinyawiwatkul)
- *Advances in Dairy Ingredients* (Geoffrey W. Smithers and Mary Ann Augustin)
- *Bioactive Proteins and Peptides as Functional Foods and Nutraceuticals* (Yoshinori Mine, Eunice Li-Chan, and Bo Jiang)
- *Biofilms in the Food Environment* (Hans P. Blaschek, Hua H. Wang, and Meredith E. Agle)
- *Calorimetry in Food Processing: Analysis and Design of Food Systems* (Gönül Kaletunç)
- *Coffee: Emerging Health Effects and Disease Prevention* (YiFang Chu)
- *Food Carbohydrate Chemistry* (Ronald E. Wrolstad)
- *Food Ingredients for the Global Market* (Yao-Wen Huang and Claire L. Kruger)
- *Food Irradiation Research and Technology* (Christopher H. Sommers and Xuetong Fan)
- *Foodborne Pathogens in the Food Processing Environment: Sources, Detection and Control* (Sadhana Ravishankar, Vijay K. Juneja, and Divya Jaroni)
- *High Pressure Processing of Foods* (Christopher J. Doona and Florence E. Feeherry)
- *Hydrocolloids in Food Processing* (Thomas R. Laaman)
- *Improving Import Food Safety* (Wayne C. Ellefson, Lorna Zach, and Darryl Sullivan)
- *Microbial Safety of Fresh Produce* (Xuetong Fan, Brendan A. Niemira, Christopher J. Doona, Florence E. Feeherry, and Robert B. Gravani)
- *Microbiology and Technology of Fermented Foods* (Robert W. Hutkins)
- *Multiphysics Simulation of Emerging Food Processing Technologies* (Kai Knoerzer, Pablo Juliano, Peter Roupas, and Cornelis Versteeg)
- *Multivariate and Probabilistic Analyses of Sensory Science Problems* (Jean-François Meullenet, Rui Xiong, and Christopher J. Findlay)
- *Nanoscience and Nanotechnology in Food Systems* (Hongda Chen)
- *Natural Food Flavors and Colorants* (Mathew Attokaran)
- *Nondestructive Testing of Food Quality* (Joseph Irudayaraj and Christoph Reh)
- *Nondigestible Carbohydrates and Digestive Health* (Teresa M. Paeschke and William R. Aimutis)
- *Nonthermal Processing Technologies for Food* (Howard Q. Zhang, Gustavo V. Barbosa-Cànovas, V.M. Balasubramaniam, C. Patrick Dunne, Daniel F. Farkas, and James T.C. Yuan)
- *Nutraceuticals, Glycemic Health and Type 2 Diabetes* (Vijai K. Pasupuleti and James W. Anderson)
- *Organic Meat Production and Processing* (Steven C. Ricke, Michael G. Johnson, and Corliss A. O'Bryan)
- *Packaging for Nonthermal Processing of Food* (Jung H. Han)
- *Preharvest and Postharvest Food Safety: Contemporary Issues and Future Directions* (Ross C. Beier, Suresh D. Pillai, and Timothy D. Phillips, Editors; Richard L. Ziprin, Associate Editor)
- *Processing and Nutrition of Fats and Oils* (Ernesto M. Hernandez and Afaf Kamal-Eldin)
- *Processing Organic Foods for the Global Market* (Gwendolyn V. Wyard, Anne Plotto, Jessica Walden, and Kathryn Schuett)
- *Regulation of Functional Foods and Nutraceuticals: A Global Perspective* (Clare M. Hasler)
- *Resistant Starch: Sources, Applications and Health Benefits* (Yong-Cheng Shi and Clodualdo Maningat)
- *Sensory and Consumer Research in Food Product Design and Development* (Howard R. Moskowitz, Jacqueline H. Beckley, and Anna V.A. Resurreccion)
- *Sustainability in the Food Industry* (Cheryl J. Baldwin)
- *Thermal Processing of Foods: Control and Automation* (K.P. Sandeep)
- *Trait-Modified Oils in Foods* (Frank T. Orthoefer and Gary R. List)
- *Water Activity in Foods: Fundamentals and Applications* (Gustavo V. Barbosa-Cànovas, Anthony J. Fontana Jr., Shelly J. Schmidt, and Theodore P. Labuza)
- *Whey Processing, Functionality and Health Benefits* (Charles I. Onwulata and Peter J. Huth)

# CONTENTS

# CONTRIBUTORS

**Murat O. Balaban**
Professor, University of Alaska, Fairbanks, AK;
e-mail: mob@sfos.uaf.edu

**Dorin Boldor**
Assistant Professor, Biological and Agricultural Engineering Department, Louisiana State University, Baton Rouge, LA;
e-mail: dboldor@agcenter.lsu.edu

**David Bresnahan**
Research Principal, Kraft Foods, Inc., Glenview, IL;
e-mail: dbresnahan@kraft.com

**Ray Carroll**
Director of process safety, Campbell Soup Co., Campden, NJ;
e-mail: raymond_carroll@campbellsoup.com

**Antoine Cazier**
Senior Project Manager, Centre Technique de la Conservation des Produits Agricoles (CTCPA), Dury, France;
e-mail: acazier80@hotmail.com

**I. Figueroa**
Graduate Student, University of Pittsburgh, PA;
e-mail: imf3@pitt.edu

**Jean Larousse**
Former Director of Centre Technique de la Conservation des Produits Agricoles (CTCPA), Dury, France;
e-mail: jean.larousse@wanadoo.fr

**Cristina Sabliov**
Assistant Professor, Biological and Agricultural Engineering Department, Louisiana State University, Baton Rouge, LA;
e-mail: csabliov@agcenter.lsu.edu

**K.P. Sandeep**
Professor, Department of Food, Bioprocessing and Nutrition Sciences, North Carolina State University, Raleigh, NC;
e-mail: kp_sandeep@ncsu.edu

**Ricardo Simpson**
Professor, Departamento de Procesos Químicos, Biotecnológicos, y Ambientales; Universidad Técnica Federico Santa María, Valparaíso, Chile; e-mail: ricardo.simpson@usm.cl

**Arthur A. Teixeira**
Professor, Department of Agricultural and Biological Engineering, University of Florida, Gainesville, FL; e-mail: atex@ufl.edu

**François Zuber**
Deputy Scientific Manager, Centre Technique de la Conservation des Produits Agricoles (CTCPA), Dury, France;
e-mail: fzuber@ctcpa.org

# Chapter 1

# INTRODUCTION

*K.P. Sandeep*

Thermal processing of foods in one form or the other has been in place since the 1900s. Although the fundamental principles remain the same, there have been numerous improvements in the control and automation of thermal processes. The various chapters in this book provide an insight into the details of the control and automation processes and details involved for different thermal processes. In order to fully understand and appreciate these details, it is important to have an understanding of the improvements that have taken place in equipment design (novel heat exchangers), process specifications (lower tolerances), product formulations (new types of ingredients), enhancement of quality (by decreasing the extent of overprocessing), and process safety requirements (identification and control of critical parameters in a process). All these are based on the fundamental and practical understanding of various topics. A brief summary of these topics is presented in this chapter.

## 1.1. Composition and classification of foods

Processed foods consist of carbohydrates (C, H, and O), proteins (C, H, O, and N), fats (usually glycerol and three fatty acids), vitamins, enzymes, flavoring agents, coloring agents, thickening agents, antioxidants, pigments, emulsifiers, preservatives, acidulants, chelating agents, and replacements for salt, fat, and sugar. Some of these are naturally present in the food, while some others are added for

*Thermal Processing of Foods: Control and Automation.* Edited by K.P. Sandeep

achieving specific functionality. Addition of different ingredients to a food product may have an effect on the stability, functionality, or properties of the food and have to thus be added in precise and predetermined quantities. During a thermal process, these constituents of a food product may undergo changes, resulting in changes in the properties, quality, and physical appearance of the food product as a whole, some of which may not be desirable. Thus, it is important to minimize the extent of thermal process a food receives.

Foods are generally classified as low acid if their equilibrium pH is greater than or equal to 4.6 and acid if their equilibrium pH is less than 4.6. The choice in the pH value of 4.6 arises from the fact that it has been documented by various researchers that the most heat-resistant pathogenic organism of concern in foods, *Clostridium botulinum*, does not grow at pH values below 4.6. Low-acid foods that have a water activity of 0.8 or higher and are stored under anaerobic and nonrefrigerated conditions have to undergo a very severe thermal process to ensure adequate reduction in the probability of survival of *C. botulinum*, in order to render the product commercially sterile. Acid products, on the other hand, need to be subjected to a much milder heat treatment as the target organisms are usually molds and yeasts. Thus, it is important to know if the product under consideration for thermal processing belongs to the low-acid or acid category.

## 1.2. Preservation of foods

A food can be preserved (under refrigerated or nonrefrigerated conditions) by several methods. Some of the commonly used techniques include the lowering of its water activity (by dehydration, cooling, or addition of salt/sugar), removal of air/oxygen, fermentation, and removal/inhibition/inactivation of microorganisms. Commercial and large-scale operations associated with preservation of foods by inactivating microorganisms usually include thermal processing. Foods meant to be refrigerated are generally subjected to a pasteurization treatment, while foods meant to be shelf-stable are subjected to retorting, hot-filling, or an aseptic process. The quality of the ingredients used, the degree of thermal treatment, the packaging used, and the storage conditions affect the shelf life of the foods.

## 1.3. Properties of foods

The properties of importance in thermal processing of foods are the physical (density, viscosity, and glass transition temperature), thermal (thermal conductivity and specific heat for conventional heating), electrical (electrical conductivity for ohmic heating), and dielectric (dielectric constant and loss factor for microwave and radiofrequency heating). Some of the other product characteristics to be considered are the shape, size, water activity, ionic strength, denaturation of protein, and gelatinization of starch. Some of the product system characteristics of importance are the heat transfer coefficients, pressure drop, and extent of fouling. Many of these properties are dependent on a variety of factors, but most importantly on temperature. Several empirical correlations exist to determine the properties of many foods as a function of their composition and temperature.

## 1.4. Heating mechanisms

Numerous methods exist for thermal processing of foods. Some of these techniques include the use of steam injection, steam infusion, tubular heat exchangers, shell and tube heat exchangers, plate heat exchangers, scraped surface heat exchangers, extruders, ohmic heaters, infrared heaters, radiofrequency heaters, microwave heaters, and variations/combinations of these. The choice of the heating mechanism is based on several factors including the nature of the product (inviscid, viscous, particulate, etc.), properties of the product (thermal, electrical, and dielectric), floor space available, need for regeneration, need or acceptability of moisture addition/removal, nature heating required (surface versus volumetric), ease of cleaning, and of course, cost (capital and operating).

## 1.5. Microorganisms and their kinetics

Microorganisms are classified as aerobes and anaerobes (either facultative or obligate) depending on their need for the presence or absence, respectively, of oxygen, for their growth. They may also be classified as psychrotrophs (grow under refrigerated conditions), mesophiles (grow under ambient/warehouse conditions), or thermophiles (grow under temperatures encountered in deserts) and can be obligate or facultative. Thus, on the basis of the package environment (presence or absence of oxygen/air) and storage temperature, the organisms that can proliferate vary. Thus, these factors, along with the other important factors (pH and water activity), form the basis for the determination of the target organism for processing any product.

The inactivation of most bacteria (at a constant temperature) usually follows the first-order kinetics reaction described by the following equation:

$$N = N_0 10^{-t/D_{\mathrm{T}}} \tag{1.1}$$

where $N_0$ is the initial microbial count, $N$ is the final microbial count, $t$ is the time for which a constant temperature is applied, and $D_{\mathrm{T}}$ is the decimal reduction time.

The effect of temperature on the heat resistance of microorganisms is generally described by the D-z model given by the following expression:

$$D_{\mathrm{T}} = D_{\mathrm{ref}} 10^{(T_{\mathrm{ref}} - T)/z} \tag{1.2}$$

where $T_{\mathrm{ref}}$ and $D_{\mathrm{ref}}$ are the reference temperature and the decimal reduction time at the reference temperature, respectively, and $z$ is the temperature change required for an order of magnitude change in the decimal reduction time.

An alternate and more fundamental approach describing the heat resistance of microorganisms as a function of temperature is the Arrhenius kinetics approach and is given by the following equation:

$$k = A\mathrm{e}^{-E_{\mathrm{a}}/RT} \tag{1.3}$$

where $k$ is the reaction rate, $A$ is the collision number (or the frequency factor), and $E_{\mathrm{a}}$ is the activation energy.

Due to the simplicity of the D-z model, it is the preferred model for use in the food industry to describe the effect of temperature on the inactivation of microorganisms. It should be noted that the link between the D-z model and the Arrhenius model is provided by the following equation:

$$E_a = \frac{2.303R(T)(T_{ref})}{z} \tag{1.4}$$

## 1.6. Process safety and product quality

Once the target microorganism is identified and the kinetic parameters ($D$ and $z$ values) of the organism are determined, a thermal process (time and temperature) is then designed to reduce the population of the target microorganism to an acceptable level (that level depends on the product characteristics process categories discussed in the preceding sections). Even for a constant temperature process, it should be noted that several combinations of time ($t$) and temperature ($T$) can result in identical levels of inactivation of microorganisms. The $F$ value, described by the following equation, is used to describe these combinations:

$$F = 10^{T-T_{ref}/z}t = D_{ref} \log \frac{N_0}{N} \tag{1.5}$$

Nonisothermal process temperatures are handled by integrating the above equation with temperature as a function of time.

For both isothermal and nonisothermal temperatures, an $F$ value can be computed for any process, based on the above equation. This value has to be equal to or greater than the predetermined $F$ value for the process to be safe. It is easy to see that the minimum required $F$ value can be achieved by increasing the process time or temperature. However, it should also be noted that different quality and nutritional attributes of the food will be lost at different rates and to different degrees at different combinations of time and temperature. Thus, a process optimization has to be conducted to ensure food safety and maximize product quality. The cook value ($C$), given by the following equation, is used to determine the critical quality attribute of concern within a food product:

$$C = 10^{(T-T_{ref})/z_c}t \tag{1.6}$$

The above equation describing the cook value ($C$) is very similar to the equation for $F$ value (equation (1.5)). The main differences between the two equations are the choice of the reference temperature (generally, $T_{ref} = 121.1°C$ for computing the $F$ value and $T_{ref} = 100°C$ for computing the $C$ value) and the magnitudes of $z$ and $z_c$ (generally, $z = 10°C$ and $z_c$ is much greater than 10°C).

The process of optimization involves ensuring food safety by making sure that the $F$ value obtained using equation (1.5) is at least the minimum value required for that type of product and at the same time minimizing the $C$ value of the critical quality attribute obtained using equation (1.6). For the case of $z_c$ greater than $z$, this optimization process results in recommending the use of higher temperatures for short times.

## 1.7. Concluding remarks

A thorough knowledge of the above-described topics is important to fully understand the control and automation of various thermal processes. The chapters that follow discuss details starting from techniques of process controls and build up to process control of retorting and aseptic processing, strategies to correct deviant thermal processes, optimization of thermal processes, and control and modeling of continuous flow microwave processing.

# Chapter 2

# ELEMENTS, MODES, TECHNIQUES, AND DESIGN OF PROCESS CONTROL FOR THERMAL PROCESSES

*David Bresnahan*

## 2.1. Introduction

Thermal processes are used to develop the product quality and food safety aspects of many food products. Control of the process parameters is therefore critical to the ability to produce a quality product while ensuring product safety. Often the thermal process effects on the product quality attributes are inverse to the effects on product safety attributes, and therefore precise control becomes even more important.

One definition of *process control* could be "the measurement and control of process variables to achieve the desired product attributes." Again, the paramount process attribute in many thermal processes is food safety. Proper design, implementation, and validation of the system are key to achieving this result.

Automatic control provides greater consistency of operation, reduced production costs, and improved safety. A process that is vulnerable to upsets is going to have a more consistent output if the process variables are adjusted constantly by an automatic control system. Human variability can be taken out of an operation with a properly implemented automatic control system.

*Thermal Processing of Foods: Control and Automation*. Edited by K.P. Sandeep

Improved consistency of operation can produce products with attributes closer to specification targets, thereby increasing overall quality. Closer control can also lead to less out-of-specification product and help ensure operation within the critical food safety limits, and therefore increase productivity.

Process control comes in two distinct formats, discrete or *digital* and continuous or *analog* controls. These two modes are often intertwined in the overall system. The combination of the two forms is usually very important in ensuring that only safe and acceptable quality products reach the consumer.

## 2.2. The process model

A process model depicting negative *feedback control* is shown in Figure 2.1. The process variable to be controlled is measured. The process measurement is compared to a set point to generate an error signal. The error signal is used by an algorithm to determine the control response. The control response is then used to manipulate a final control element that affects the control variable and the loop is repeated.

An example of negative feedback control is a typical temperature control loop whereby a fluid is heated as it passes through a steam heat exchanger. The fluid temperature is the control parameter. A temperature *sensing element* (sensor) is used as the measurement device. Judgment of whether the temperature is too high or low is

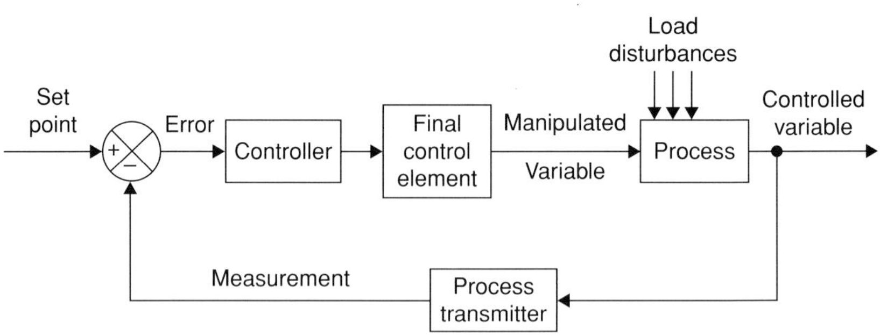


**Figure 2.1.** Process model.

made by the *controller* by comparing the measured value to a preset *set point*. The steam valve is used to make appropriate adjustments. If the fluid is too hot then the controller sends a signal to adjust the steam valve toward the closed position; thus the concept of negative feedback control. A positive error requires a negative response for correction. When the deviation of the fluid temperature from the set point is large, the controller adjustments are large. As the desired set point is approached, the controller makes finer and finer adjustments.

## 2.3. Automatic control loop elements

Figure 2.1 indicates the information flow in a feedback control loop configuration. The elements within the loop can vary but are often similar.

The process variable is detected with a sensing element or *transducer*. A transducer is a device that produces an output in some relationship to the measured parameter. Very often the transducer signal is fed to a *transmitter* that changes the transducer signal to a standardized signal and sends it on to the *controller*. The controller determines the control response and then sets the *controller output*. The controller output is a standardized signal that goes to another transducer that converts this signal to a proportional signal that drives the *control element*.

For example, a temperature control loop might consist of the following elements. A resistance temperature detector (RTD) is the transducer used to measure the product temperature. This device changes resistance with temperature. A transmitter then produces a 4–20 mA signal in proportion to the calibrated range of resistances. The controller reads the 4–20 mA signal and interprets this in engineering units, compares it to the set point and generates a control response of 0 100%. The control signal is sent out as another 4–20 mA signal in direct proportion to the control response. The control signal goes to another transmitter (a current to pneumatic converter) that outputs a pneumatic signal of 3–15 psig in direct proportion to the inlet control signal of 4–20 mA. The pneumatic signal of 3–15 psig then proportionally drives a control valve from 0% to 100% open.

Table 2.1 lists some common measurement devices used in thermal processes. Careful consideration needs to be given when selecting devices for a particular application. Accuracy and repeatability are important criteria. For instance, in a retort where a mercury-in-glass (MIG) thermometer is the reference device, the control and recording instruments should be able to reliably provide readings that are very close to those of the standard. This will allow the system to operate much closer to the critical limit providing adequate food safety while reducing the impact on product quality.

A scheduled calibration program is important for maintaining the integrity of the system. The sensors that measure the critical variables are generally calibrated or have their calibrations checked on a more frequent basis than those instruments that measure noncritical parameters.

Redundancy may also be considered for some critical variables. An example would be using an RTD probe that has two elements in the same housing. The transmitter then compares the two RTD signals to make sure they are within a specific tolerance to help ensure the system is accurate and working properly. This might be used in such critical applications as the end of a hold tube in an aseptic process or as the temperature control element in a retort.

For sensors in contact with the product it is required that the contact surfaces be constructed of approved food contact materials. All liquid applications do not require 3A approval, but this certification indicates that this sensor can be used in clean-in-place (CIP) applications without much extra consideration by the design engineer. Sensors in a process that will be CIPed should be mounted to minimize any dead volume and be self-draining.

Sensor installation is important for proper functioning. A temperature sensor should make good contact with the material being measured. Flow sensors often require certain lengths of straight piping runs up- and downstream of the flow element. Some sensors are vibration sensitive, while others are susceptible to electrical noise.

Just as care needs to be taken in selecting a sensing element, the sizing of the final control element (typically a valve or pump) is also critical for the proper functioning of a control loop. If the response of the final control element is too large or small in proportion to

**Table 2.1.** Common sensors list for thermal processes

| Parameter | Sensor | Application range | Comment/special considerations |
|---|---|---|---|
| Temperature | Resistance temperature detector (RTD) | −430 to 1200°F | 100 Ω Platinum most widely used. Good accuracy—3 or 4 wire models recommended for improved accuracy |
| Temperature | Thermocouple | | Moderate accuracy—long runs of thermocouple wire not recommended due to low signal level |
| | Type J | −320 to 1400°F | |
| | Type T | −310 to 750°F | |
| | Type K | −310 to 2500°F | |
| Volumetric flow | Magnetic | Down to 0.01 gpm | Some minimal conductivity required. Good accuracy for most control applications |
| Volumetric flow | Vortex shedding | Down to 10 gpm | High Reynolds required. Can be used for steam. Often used in CIP circuits |
| Mass flow | Coriolis | Down to 0.1 lb/min | Highly accurate, pressure drop can be a concern. Density also measured independently. Good for metering applications (e.g., batching) |
| Mass flow | Heat loss | Down to 0.5 cc/min | Commonly used for gasses |
| Pressure | Strain gauge | Down to 2 psi | High temperature applications (>250°F), usually require isolation mounting |

the control correction required, then it will be difficult to achieve consistent accurate control of the process variable.

The loop communication used in the example above uses standard signals that are very common. However, digital networks offer an alternative that can be more cost effective to install and maintain. Installation of the instruments on a digital network can require just one set of wires to connect the devices in series instead of one set of wires per instrument. Maintenance is enhanced because of the inherent intelligence in the device and network controller that can provide information on when a device is starting to fail, and if it does fail it can help in quickly locating the failed device.

## 2.4. Process dynamics

Processes are often in need of adjustment due to many factors. The output of a process may have to fluctuate to match the needs of downstream operations and efficiencies. The process demands will also vary depending on the current phase of the process, for instance: heating, holding, and cooling of a batch retort, and the sterilization, product startup, continuous processing, shutdown, and cleaning for a pasteurizer. Even during one phase of a process where throughput is being held constant, there can be various upsets such as changes in utility supplies that subsequently require a compensating adjustment.

There are process dynamics that will delay the response time of a system. Two delaying characteristics are lag and dead time.

Capacitance is the ability of a system component to store energy. At the same time system components can impede the rate of energy transfer. The capacitance and resistance of energy transfer result in control system response lags. If in the heating of a batch retort the steam flow is suddenly increased, the system lags due to the amounts of materials that need to be heated (the vessel and its contents) and the rate of heat transfer as determined by the cumulative resistances.

*Dead time* is a delay in the response of a system usually due to a transport phenomenon. As the product in a pasteurizer or aseptic sterilizer passes out of the final heat exchanger through the hold tube to a temperature detector, part of the delay in the temperature rise detection will be due to the time it takes for the heated fluid to reach the hold tube temperature probe. To eliminate most of the effects of this delay on the controller response, a temperature probe for control is generally placed just at the exit of the heater.

## 2.5. Modes of control

### *2.5.1. On/off control*

The *on/off control* algorithm is trivial in terms of its mathematical expression; however, it is applicable to many control situations. On/off control works by completely turning the control element on or off in response to a change in sign of the error. For instance, when heating a batch retort, once the vessel temperature crosses the set point, the valve on the heating medium supply is turned off. After the vessel subsequently cools back down below the set point, the heating media supply valve is turned back on.

For the initial heating to set point, the on/off control will generally result in an overshoot because the heating valve does not shut off until the set point is achieved. In some applications, such as retorts, this can be desirable in order to get the usually slower responding MIG up to the cook temperature sooner. For many other processes this overshoot is not desirable as it can lead to reduced product quality.

For more constant demand systems, such as a retort during the cook cycle or a continuous flow heat exchanger, on/off control will approximately maintain the average temperature as the set point; however, there can be substantial excursions above and below the set point depending on the dynamics of the system involved. This type of control in continuous flow system is generally not good enough for the critical parameters.

### *2.5.2. Proportional, integral, and derivative control*

For continuous systems and for a lot of parameters in a batch system, the most popular type of controller is a *proportional, integral, and derivative (PID) controller*. This control algorithm calculates the amount of control action to take from 0 to 100%. This generally feeds to a proportioning valve that opens or closes the corresponding amount. When combined with an on/off control valve, the valve would open for the control signal proportion of a given time frame. For instance, if the controller calculates a control response of 60%, a direct acting proportional valve would open to the 60% position. With the same signal, an on/off valve might stay open for 6 seconds out of every 10.

The three modes of the PID controller contribute to the calculation of the control signal in different ways. The *proportional* portion of the equation is a response in proportion to the error (the difference between the set point and actual measurement). Using this type of control alone will result in a system where there is always an offset of the measured signal to the set point. The *integral* portion of the controller provides controller output information based on the error accumulated over time. Adding this mode of control will get rid of the offset from a proportional-only controller, but will give some degree of overshoot as the set point is approached. The *derivative* portion of the control action is derived from the rate of change of the error. Adding this mode of control will allow for the reduction in the amount of overshoot; however, this mode should only be employed on control loops where some lag or dead time exists.

The PID modes can be used in many different combinations. Table 2.2 shows common combinations for some commonly controlled variables. It is even possible to combine on/off control with derivative control in order to minimize overshoot with a simple "combined" control algorithm.

On/off and PID control are not the only control algorithms, but they are the most common. Many enhancements are available that may be useful for certain control problems. Some of these enhancements (such as model-based control) can be used as supervisory controllers, providing set points to standard PID controllers.

**Table 2.2.** PID parameter applications

| Controlled variable | Proportional | Integral | Derivative |
|---|---|---|---|
| Temperature | Medium to high | Low to medium | Low to medium |
| Flow | Medium to high | Medium to high | None to low |
| Pressure | High | Medium to high | None |
| Level | Low to high | None | None to low |

## 2.6. Controller tuning

A PID controller is a very useful tool. However, proper setting of the PID parameters is required to get the desired results. The type of response desired will vary depending on the application. In some applications the desire may be to get to the set point as fast as possible. Batch retort heating could be an example of this. In other systems it may be very important to minimize the amount of overshoot or undershoot. The "cook" phase of a batch retort or a continuous flow pasteurizer that must maintain the heated product above a critical limit are examples where this type of fluctuation cannot be tolerated.

Sometimes a combination of different characteristics is desired such that either compromises are made to get some of each characteristic or changes in the control loop are made for different circumstances. For instance, a batch retort might have one set of tuning parameters to get the system to temperature quickly and another set of parameters for the low load "cook" phase.

Even when a controller is tuned, the response cannot be expected to be the same over 100% of the range of the controlled variable and the various upsets. For instance, the gain of a heating loop at high flow rate may be fairly low, that is, the change in product temperature for every 1% valve opening is low. When the flow is slowed, the change in product temperature for every 1% of valve opening will be greater and thus the loop gain is higher. In addition, the slower flow rate will increase the dead time in the detection of a temperature change. This increased response and increased dead time may cause poor or even unstable control response.

Some components of the control loop are not linear in their response and thus the gain of the loop will change in these areas. Control valves can have many different responses. If a valve is grossly oversized, the upper range of the controlled variable may be reached with a small valve opening. Control becomes difficult because a large portion of the control response is useless.

One common type of control response is the quarter decay response as first described by Ziegler and Nichols (1942). Quarter decay is defined as having the area under the response curve reduced by one-fourth for each subsequent excursion on the same side of the set point. This type of response is designed to provide a fast response while also keeping the total error small.

Again this may not be the most desired response, particularly if undershoot or overshoot cannot be tolerated. Other tuning objectives and more detail on tuning methodologies can be found in McMillan (1994), Corripio (1990), and Liptak and Venczel (1985a).

Since the controller response can vary over the range of control, the optimal tuning parameters will also vary. The objective of tuning is to get the best response in the normal operation range while achieving acceptable response in other areas. Often, experience and considerable patience are required to get acceptable performance from a control loop. Applying more sophisticated control techniques can help to overcome the problems of some loops. However, from an operation and maintenance perspective, control schemes should be kept as simple as possible.

Controllers with self-tuning capability are available. This feature can be a great aid in the start-up and long-term operation of a process. However, their mode of operation must be understood and applied properly in order to get value from the self-tuning ability.

## 2.7. Control loop troubleshooting

Proper design of a control loop with all of its components is very important in getting a loop to perform well. As with most things, periodic loss in performance may occur.

Very often, when a control loop is performing poorly, the first reaction is to adjust tuning parameters to fix the problem. For a previously working loop, adjusting the tuning may make it perform better, but most likely it will only be fixing a symptom without finding the cause.

Before changing the tuning on a loop that starts to perform poorly, a systematic investigation of each loop component needs to take place (Valentis *et al.*, 1997). Very often one of the loop components is not working properly, causing the issue with the previously operating system. Some examples of items that may be causing the issue include, sensor not working properly (e.g., loose connection), utility not under control (e.g., steam regulator not working or chilled water supply temperature fluctuating or plug in a utility line), steam trap not functioning, control valve stuck, leak in pneumatic line to control valve, current-to-pneumatic converter not working, utility supply valve not opening properly, and the heat exchanger being fouled.

Only after being satisfied that all the loop components are working properly should retuning the loop be considered, keeping in mind that this loop was working once and therefore something must have changed. Care should be taken while tuning a loop to make sure the new tuning can handle the range of load conditions that will be encountered. For instance, for heating and cooling control loops, the performance should be checked at the high and low flow rates that are to be required.

## 2.8. Process and instrument drawing (P&ID) symbology

The representation of control techniques/schemes is generally done on process and instrument drawings (P&IDs). These drawings show the relationship of the sensing elements to the process equipment, although not to scale. The connections of the different control elements are also shown. P&IDs are used for design, installation, and troubleshooting of control schemes and as such there can be different versions with different levels of detail.

Figure 2.2 shows a simple control scheme with explanations for the symbols used in this text. The ANSI/ISA-S5.1-1984 *Instrumentation Symbols and Identification* standard has an extensive list of symbols. This resource should be primary when developing P&IDs in order to easily convey the design to the many parties involved.

## 2.9. Control techniques

There are many possible combinations of control elements. The different combinations are called control techniques. A brief discussion of the most common techniques used in thermal processes is given below.

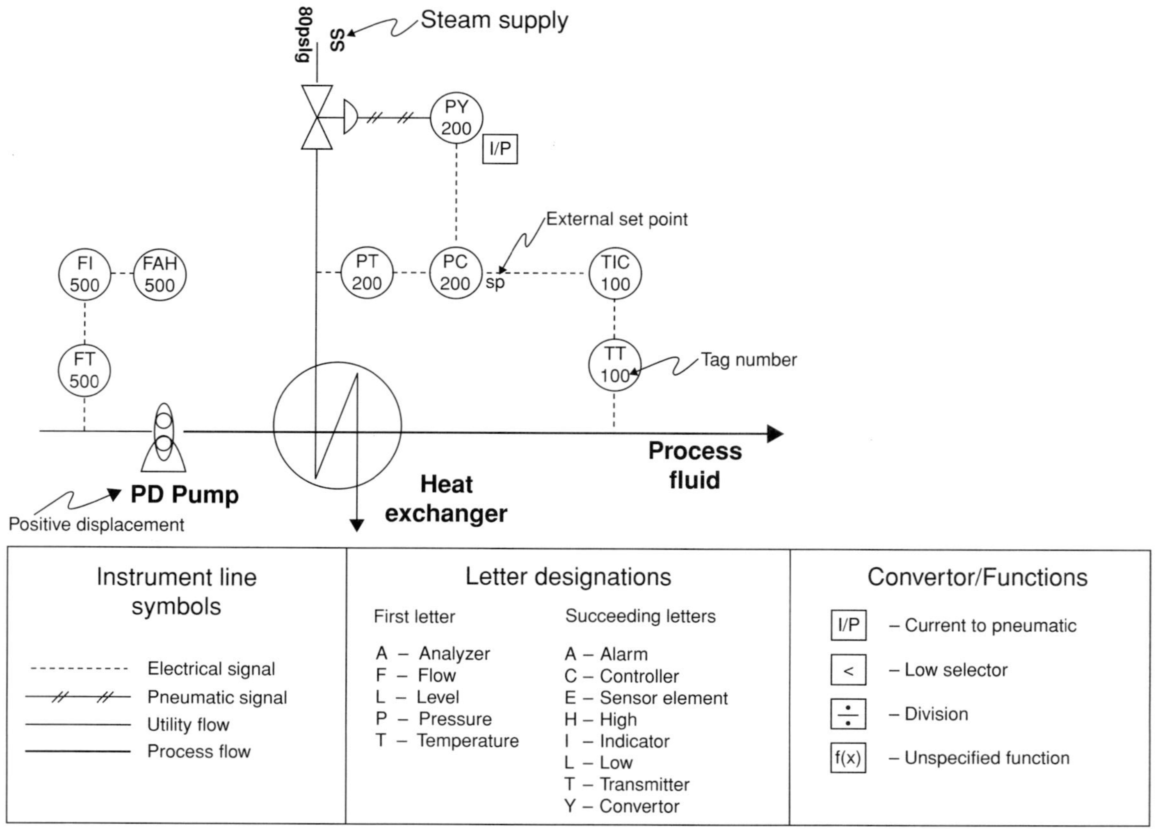


**Figure 2.2.** Process and instrument diagram.

### *2.9.1. Negative feedback control*

Negative feedback control is the most common control technique. The term feedback comes from the fact that the controlled variable is measured after being influenced by the final control element. This does not mean that the controlled variable transducer is always located downstream of the final control element. Rather, the feedback refers to the flow of information in a backward loop as shown in Figure 2.1. The feedback information is used by the controller to attenuate the effects of disturbances and to bring the process variable back to set point. The negative refers to the change in sign that information must make in the loop to bring the error toward a value of zero. Figure 2.3 shows schematics of three feedback control loops.

Feedback control is simple to implement and therefore is widely used. One drawback is that the controller is always reacting to events that have already happened. The sensor measures the effects of an upset and then responds. For control loops with long lags or dead times, this type of control may not be able to handle large upsets.

Feedback control is oscillatory in nature. Improper setting of the controller gain can lead to instability. Nonlinearities in the control loop components will lead to an inconsistent response across the total range of control. Feedback control loops often have a lag between when a correction is made and when the effects of that correction are measured. This can cause overshoot of the set point or make the control unstable due to constant over correction.

Even with some disadvantages, feedback control is often the heart of a control system design. Combining other techniques with feedback control can usually overcome the deficiencies while keeping the overall system easy to understand and maintain.

The controllers in Figure 2.3 can have many disturbances; the steam supply pressure, cooling medium supply temperature, and product flow rate are some examples. Appropriate use of environmental control can greatly improve the performance of other control loops. An example of this would be the addition of a steam supply regulator may help reduce the variability in the heating controller. While this removes some of the disturbances, others such as heat exchanger fouling and room temperature and humidity may still affect the heating control.

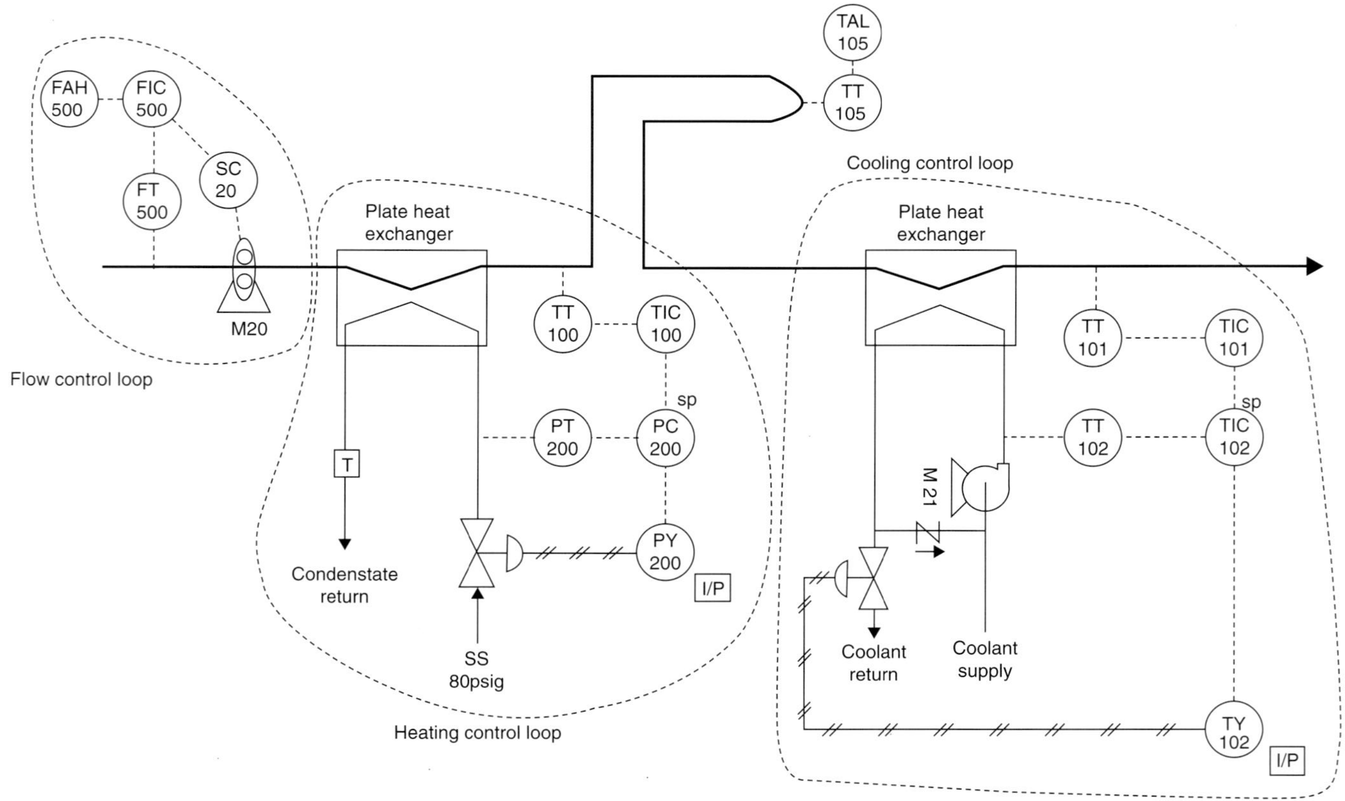


**Figure 2.3.** Control loops.

Not all disturbance variables can practically be held constant. In the temperature loops, if the product flow rate needs to change to meet downstream demands, the rate of change of the flow rate set point should be limited in order that the heating and cooling control loops can keep up with the changing demand.

### *2.9.2. Cascade control*

*Cascade control* is another technique that can be used to improve the performance of some control loops and to help overcome some system disturbances. The control is split into two parts; a secondary (inner) loop and a primary (outer) loop. The primary loop output is used as the set point for the secondary loop.

Both the heating and cooling control loops in Figure 2.3 are set up as cascade control loops. The heating loop has TC-100 as the primary loop and steam pressure controller PC-200 as the secondary. In the cooling control, TC-101 is the primary loop, while TC-102 controlling the temperature of the recirculating cooling media is the secondary loop.

The secondary loop can take care of upsets in the manipulated variable before they impact the primary variable. With the manipulated variable already under feedback control, the primary control loop may be more linear. The primary loop controller adjusts a set point that is more linearly related to the primary variable than is the position of the final control element. The primary variable speed of response will be improved through the application of cascade control if a lag exists in the secondary control loop.

The response speed of the secondary loop needs to be faster than that of the primary loop. Otherwise the primary loop will always be making set point changes to the secondary loop without enough time passing for the effects of those changes to be realized. Loops interacting in this fashion will result in constant oscillations of the primary loop. Oscillation elimination is accomplished by making the secondary loop faster by a factor of three or more than the primary loop.

Note that making the secondary loop three times faster than the primary loop does not mean taking the tuning parameters of the primary loop and multiplying or dividing them by three. Rather, this implies that the correction of a given error by the secondary loop will be three times faster than the correction to the same percent error in the primary loop. The tuning constants for the two loops will generally bear no relationship to each other due to the differences in the individual loop gains.

### 2.9.3. Interlocks

*Interlocks* are the marrying together of digital and analog control to have the system react properly to changing process conditions. An interlock is the recognition of the system of the existence of a certain operational state or event and a reaction to that state or event. Interlock signals can come from physical switches such as a high level switch in a tank, software switches where the control computer is comparing an input signal from a sensor against a limit, or the general recognition of the control computer of what step the process is in and the conditions needed to maintain that step or transition to the next step.

Often interlocks are used to initiate an event or series of events that prevent an unsafe or otherwise undesirable process state from being reached. In the example of high temperature short time (HTST) processing systems, high flow and low temperature alarms are used to divert underprocessed product away from downstream operations, preventing it from ultimately reaching the consumer.

Indeed all the critical limits of a process that affect the operational safety and/or safety of the product to the consumer should have interlocks to prevent the undesired results. A record of the activation of the critical factor interlock should be made by the system in order to help in the verification of the production of safe product.

The operational status of the process can be used to drive other interlocks. If a line is idle then an interlock can be used to put control loops in manual and force the outputs to the closed or off positions. This can act as a secondary safety for utility shut off when the loop is inactive. For some control loops it may be better to place them in manual and leave the output at its last position. Then, when the system restarts, the controller may not have to search as long to find the right output level.

## 2.10. Control system design

The most important consideration in control system design is to know the process that is to be controlled. Cause and effect relationships between product attributes and process variables should be known in some quantitative fashion. The limits and precision required of each process variable will determine the type of control that should be applied.

If a product has little tolerance for variation of a certain process variable, that variable will require an accurate sensor, a carefully tuned controller, and proper sizing of the final control element. Careful control of other variables affecting the critical variable may also be required.

Proper control system design should consider applying the appropriate amount of technology. Superfluous technology can be distracting. Simplicity is a key for a successful implementation. Long-term success is dependent on useful functionality for the operator and ease of maintenance. Fancy graphics may look nice for touring management but can be distracting to operators and may make updates more cumbersome. Sophisticated programming techniques may provide elegant solutions to the programmer and save a few bytes of memory, but can be confusing for the maintenance personnel. Oftentimes it is helpful to have additional instrumentation during commissioning and start up of a process line, but this adds cost and increases the ongoing calibration requirements. Analog gauges can sometimes fill this role while being less expensive and easier to maintain.

Most control is inferential, that is, the product attributes (taste, texture, appearance, microbial load reduction) are not directly measured but are controlled inferentially by controlling other easier to measure variables (e.g., temperature, flow, and pressure). Often these inferences assume certain consistencies in raw materials as well as in mixing operations. Raw material variations generally occur infrequently. Set points and other operating parameters can be adjusted to compensate for the results of the laboratory checks of raw material attributes.

More sophisticated sensors to measure product attributes such as gas chromatographs, mass spectrometers, and spectrophotometers are being offered for on line use. However, these devices require an increase in sophistication of users and increased maintenance. Cost justification for these sensors should include not only their purchase price but also any additional maintenance and service contracts.

Simplicity is an important design goal. Complexity should only be added where justified. Keeping the number of devices to a minimum reduces the number of components that can potentially fail and eases the overall maintenance requirements.

Control system design needs to consider the transitory conditions that may occur in a process. Start up, shut down, and transitions between products are examples of transitory states. Proper consideration to these states can help reduce product waste and promote better steady-state operation. For example, in heating a liquid product through a series of heaters, the process line often must change its flow rate to match the throughput of the rest of the system. If the heater set points remain fixed as the line slows, the product will be more heat abused. The system can be designed to change set points and indeed shut off/turn on heaters to keep the amount of heat abuse relatively constant throughout the operational range of the line.

The system design should consider the fail-safe mode for all the devices. For instance, the RTD transmitter on a heater should fail to a high temperature in order that the controller will force its control valve closed and therefore not scorch the material in the heater. An RTD on a hold tube should fail low such that the system does not infer that it is achieving the proper safe temperature when in fact it is unknown. Utility valves should generally shut off when the system shuts down or there is a power failure. The process line's valves default states are often in a flow path that provides pressure relief for the line.

When designing a control system, the requirements of the system need to be clearly laid out in a design document. This document should specify the variables to be controlled, the critical parameters and the reaction to the variables not meeting their limits, data recording, and reporting requirements. Detailed designs are then defined to accomplish the design goals. During the installation and commissioning of the system, the performance of the system is verified against the original design goals. Subsequent validation of the system performance, particularly against performance in handling the parameters critical to food safety should be done on a periodic basis.

For a more complete description of validation refer to NFPA (2002), *Validation Guidelines for Automated Control.*

Another consideration of the control system is that of security. Sufficient security should be present to prevent unauthorized personnel from changing values and logic that are involved with the critical control points that ensure the product's food safety.

A total process system will often have many different pieces of control hardware. Consideration for standardization across the line should be given in order that the training requirements can be reduced and the communication between controllers can be easily accomplished. The system selected needs to be able to meet the loop control, interlocks, data recording, and reporting requirements of the system.

## 2.11. Examples of control loops

### *2.11.1. Heat exchangers*

Examples of liquid process temperature controls are shown in Figure 2.3. The heating control has a steam-heated exchanger with a temperature to steam pressure cascade loop. This is a very common implementation of this type of control and it can be applied to many types of heat exchangers. The cascade is not required, but does offer the advantages of faster response and provides a view of the utility requirements that is very helpful in troubleshooting. In critical control loops such as the final heater before the hold tube, it is well worth it to have the extra security of the secondary controller to take care of utility upsets before those upset effects are seen by the primary control variable.

The cooling is accomplished with a liquid-to-liquid heat exchanger. The primary loop of the cascade may not be necessary if excess heat transfer surface area is provided and the circulating utility temperature is controlled close to that desired for the product. One thing that needs to be added to this system is differential pressure monitoring and alarming. Particularly with plate heat exchangers after a kill step, the difference between the pressure of the product (cooked) and the utility (raw) must be maintained in a positive manner. This is in case there is a leak in the plate surface. With a positive pressure on the product side, the flow will be from the cooked product to the raw utility, thus preventing any recontamination.

In the example depicted, there is a pump to circulate the cooling medium through the heat exchanger. Coolant is bled in as needed to adjust the medium temperature while the flow of coolant remains constant. This *feed-and-bleed* configuration allows for efficient and uniform heat transfer even under low load conditions.

Temperature control loops have dead time due to the transport time from the point of heat transfer to the sensor and lag due to equilibration times of the heat exchanger and the sensor. It is desirable to reduce both of these to improve control.

Dead time can be minimized by locating the sensor close to the outlet of the heat exchanger. In some cases, such as with direct steam injection heating, some minimum distance is required to allow a uniform product temperature to be reached.

Lag can be reduced by utilizing small diameter temperature probes. Maintaining high flow rates of heat transfer media in systems, as shown in the cooling loop in Figure 2.3, is important in order to maintain the heat transfer rate and to avoid variations in lag time. The high flow rate will also keep the entire heat exchanger flooded, preventing channeling and inconsistent heat transfer over the surface area.

### *2.11.2. Batch retorts*

Batch retorts are vessels that offer a lot of control challenges. First the large vessel and its contents need to heat rapidly, often along a prescribed temperature and pressure ramp, then it needs to hold temperature and pressure constant under low load conditions for the "cook" phase, and finally the whole system must cool back down, again, often at prescribed temperature and pressure ramps. The control solutions described below might not all exist on any one vessel, but these solutions can be considered across a variety of control systems besides batch retorts.

During the heating there is a lot of inertia in the system that will cause a large response by the controller. This type of control is often done with an on/off controller, since the load is so high and the speed of heating is desired to be very fast. When more precise regulation is required, such as when the heating ramp is controlled, then PID control with a regulating valve may be more appropriate.

The vessel pressure is often controlled independently with the injection of pressurized air for pressurization and an exhaust valve for depressurization. If the pressure to be controlled is just above the steam saturation temperature, then an overreaction by the depressurization can cause a loss in temperature. This will cause the temperature controller to add more steam and perhaps overpressurize the vessel, and then if the depressurization controller again overreacts, the system is caught in continuous cycle.

These interacting controllers need to be decoupled in order to provide more precise control. One way to decouple control loops is by having one controller able to control its parameter faster and more precisely than the other interacting controller. In this case, since there is virtually no lag in the pressure control, this loop would be configured to provide fast but not overcompensating control.

The pressure control itself could be interacting, if for instance two independent controllers were used, one for pressurizing and one for depressurizing. If the pressurizing controller overshoots the set point, the depressurizing controller will then react. If the depressurization undershoots the set point, then the pressurizing loop again reacts, setting up a cycle in the vessel pressure that may be increasing in amplitude as time progresses.

One way to decouple these loops is to have a single controller that controls both the air supply valve and the pressure relief valve. This loop could be set up such that the standard controller output of 4–20 mA goes to two current-to-pneumatic converters (I/P). Each I/P would only react to part of the controller output signal. The I/P for the vent valve would react to the controller output in the range of 4–12 mA, where the vent valve would be 100% open at 4 mA and closed at 12 mA. The I/P for the pressure supply valve would react to the controller output from 12 to 20 mA, where the supply valve would be closed at 12 mA and 100% open at 20 mA (Figure 2.4).

During the cook phase of a batch process, the heat load on the system is generally a lot less than during the heating and cooling phases. The large load differences may dictate some changes in the controller for more precise control during this critical phase. One way

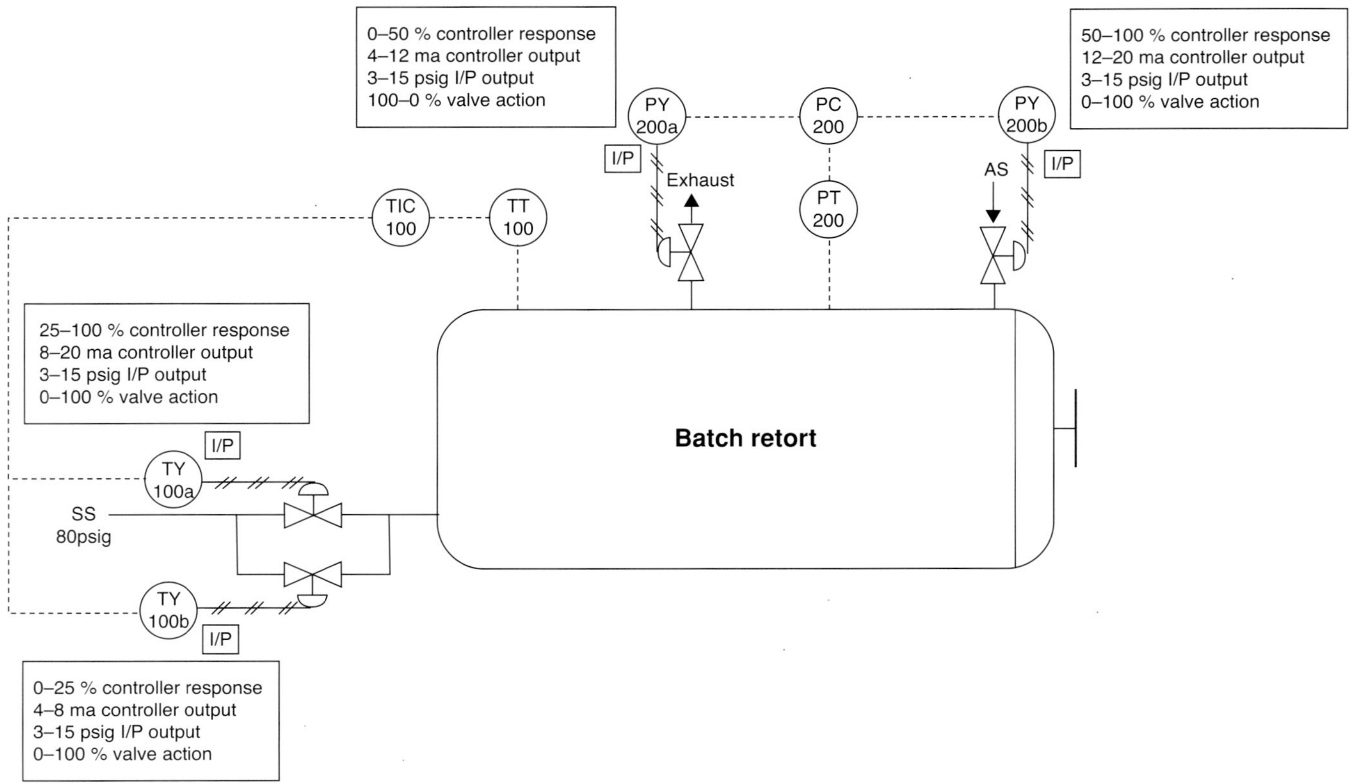


**Figure 2.4.** Control of a batch retort.

to handle the load differences could be to have the system change the PID tuning parameters of the heating controller during the cook phase. One potential issue with this is that the control valve may be oversized for this duty. Another way to compensate for the changes in duty is to provide two control valves in parallel. There are many ways to configure the control of these valves. One way is to use a single controller and the split-range I/Ps as in the pressure control example above. Here, the lower portion of the output signal, say 4–8 mA would drive the smaller control valve from 0 to 100% open. When the control signal goes above 8 mA the larger valve then begins to open, such that it would open from 0 to 100% as the control signal goes from 8 to 20 mA (Figure 2.4).

### *2.11.3. Flow*

The final control element in the flow control loop in Figure 2.3 is a variable speed pump. Flow control is often achieved with a valve downstream of a fixed rate pump (usually not a PD pump). Note that the flow sensor can be upstream of the control element. This is done because the flow disturbances caused by the control elements could affect the accuracy and stability of the flow measurement. In general, each flow sensor has particular installation requirements that should be paid careful attention for reliable flow measurement.

Even with careful attention to the flow meter installation, flow meters generally produce somewhat noisy signals. Filtering the signal can generally compensate for the noise. Filters are provided in most flow transmitters. The filter provides a type of time averaging of the output signal. While this reduces the instantaneous spikes in the signal, this is adding lag to the control circuit. As the filtering effect is increased, the lag increases and the control response will need to be slowed down.

A flow sensor may be of a smaller diameter than the rest of the piping. This may require that a bypass be installed to get sufficient flow for CIP operations.

### *2.11.4. Back pressure*

In many heating applications, the product pressure must be controlled to keep the product from vaporizing. A backpressure controller is implemented to prevent product flashing. This type of control is similar to flow control. Most often the final control element is a valve, but it can be a positive displacement pump.

The backpressure loop is often the fastest loop in a heating system. Since the backpressure may affect the flow, even when it is controlled by a pump (positive displacement pumps may still have some slippage), the flow loop will have to be tuned in a manner to prevent the two loops from interacting. This generally means that the flow loop may need to be detuned (slowed down in its response) slightly so that the flow does not change faster than the pressure controller can compensate for the effect of flow on system pressure. A consideration for flow loops that need to change flow rates during production is to ramp the flow rate set point changes so that the flow and pressure control loops can respond without major excursions from the set points instead of having step changes in set point that may cause unwanted disturbances in both loops.

## 2.12. Summary

Process control is an integral part of food thermal processes. As systems evolve, the ability to provide more precise control to improve product and package quality while maintaining operational and product safety is increasing. It is important to understand the requirements of the process to be able to design a system with the appropriate level of controls.

Even after an appropriate design is done, it is very important that the system be validated as performing as designed. Consideration for periodic verifications of system performance against critical parameters should also be given in the layout of the controls and the controlling software.

Another advantage of the evolving system technology is the ability to communicate more effectively across a network, information required by various functions such as the adherence to food safety requirements and the overall production rates.

While the functionality of control systems is evolving into many areas, the prime objective of providing control to produce safe and quality products must be kept in the forefront.

## Abbreviations

| | |
|---|---|
| CIP | Clean-in-place |
| HTST | High temperature short time |
| I/P | Current to pneumatic converter |
| mA | Milliampere |
| MIG | Mercury-in-glass |
| P&ID | Process and instrumentation diagram |
| PID | Proportional, integral, and derivative |

## Glossary

**Analog control:** Control of a variable over a continuous range of control action (e.g., 0–100% valve position)

**Bias:** A constant added to result of a calculation.

**Cascade control:** Control in which the output of one controller is the set point for another controller (ANSI/ISA, 1979).

**Clamp:** A controller setting that limits the output from going above, for a high clamp, or below, for a low clamp the set value.

**Closed loop:** A signal path that includes a forward path, a feedback path, and a summing point, and forms a closed circuit (ANSI/ISA, 1979).

**Continuous control:** Control of a variable over a continuous range of control action (e.g., 0–100% valve position).

**Control element:** A device, such as a valve or pump, that manipulates a controlled variable.

**Controller:** A device that operates automatically to regulate a controlled variable (ANSI/ISA, 1979).

**Controller output:** A signal sent from the controller in proportion to the control algorithm result.

**Dead band:** The range through an input signal may be varied, upon reversal of direction, without initiating an observable change in output signal (ANSI/ISA, 1979).

**Dead time:** The interval of time between initiation of an input change or stimulus and the start of the observable response (ANSI/ISA, 1979).

**Digital control:** Control in which devices are turned either on or off in a specific sequence or in response to certain conditions.

**Direct acting controller:** A controller in which the value of the output signal increases as the measured value increases (ANSI/ISA, 1979).

**Distributed control system (DCS):** A group of devices performing control functions connected together via a communications network.

**Derivative or rate action:** Control action in which the output is proportional to the rate of change of the input (ANSI/ISA, 1979).

**Environmental control:** Control of variables that are not the main influence of the manipulated variable.

**Error:** The algebraic difference between the ideal value (set point) and the measured signal (ANSI/ISA, 1979).

**Feed-and-bleed temperature control:** An equipment configuration where liquid heat exchange media is circulated at a high rate through the heat exchanger. New media is introduced into the loop by bleeding off some of the media from the loop in order to control the loop temperature.

**Feedback control:** Control in which a measured variable is compared to its desired value to produce an actuating error signal which is acted upon in such a way as to reduce the magnitude of the error (ANSI/ISA, 1979).

**Feed-forward control:** Control in which information concerning one or more conditions that can disturb the controlled variable is converted, outside of any feedback loop, into corrective action to minimize deviations of the controlled variable (ANSI/ISA, 1979).

**Integral or reset action:** Control action in which the output is proportional to the time integral of the input (ANSI/ISA, 1979).

**Interlocks:** Discrete signals that are used by the control system to change the mode of the process operation.

**Low selector:** A mathematical transformation of two inputs where the output is equal to the lesser of the two input values.

**Machine control:** Control in which devices are turned either on or off in a specific sequence or in response to certain conditions.

**On/off controller:** A two position controller for which one of the two discrete values is zero (ANSI/ISA, 1979).

**Open loop:** A signal path without feedback (ANSI/ISA, 1979).

**Override control:** A control scheme where one controller will take control of another controller's final control element under certain configured conditions.

**Process measurement:** The acquisition of information that establishes the magnitude of process quantities (ANSI/ISA, 1979).

**Process control:** The regulation or manipulation of variables influencing the conduct of a process in such a way as to obtain a product of desired quality and quantity in an efficient manner (ANSI/ISA, 1979).

**Programmable logic controller (PLC):** A programmable control device invented to replace electrical relay panels performing machine sequencing. Capabilities have increased substantially to include analog control as well as other features.

**Proportional and integral (PI) controller:** A controller that produces proportional plus integral (reset) control action (ANSI/ISA, 1979).

**Proportional band:** The change in input required to produce a full range change in output due to proportional control action (ANSI/ISA, 1979).

**Proportional controller:** A controller that produces proportional only control action (ANSI/ISA, 1979).

**Proportional gain:** The ratio of the change in output due to proportional control action to the change in input (ANSI/ISA, 1979).

**Proportional, integral, and derivative controller (PID):** A controller that produces proportional plus integral (reset) plus derivative (rate) control action (ANSI/ISA, 1979).

**Ratio control:** A control scheme that uses the controller to maintain a predetermined ratio between two variables (ANSI/ISA, 1979).

**Reverse acting:** A controller response in which the value of the output signal decreases as the value of the input (measured variable) increases (ANSI/ISA, 1979).

**Set point:** An input variable that sets the desired value of the controlled variable (ANSI/ISA, 1979).

**Sensing element:** The element directly responsive to the value of the measured variable (ANSI/ISA, 1979).

**Single loop controller (SLC):** A control device that has as its primary function the control of a single controlled variable at its set point.

**Time constant:** The time required to complete 63.2% of the total rise or decay for the output of a first order system forced by a step or an impulse (ANSI/ISA, 1979).

**Transducer:** An element or device that receives information in the form of one quantity and converts it to information in the form of the same or another quantity (ANSI/ISA, 1979).

**Transmitter:** A transducer that responds to a measured variable by means of a sensing element, and converts it to a standardized transmission signal that is a function only of the measured variable (ANSI/ISA, 1979).

**Variable frequency drive:** A device used to control motor speed by altering the motor's electric power frequency.

**Vena contracta:** The point along a flow stream after an orifice in which the lowest pressure is reached. This is one of the desired measuring points for an orifice plate differential pressure flow meter.

## References

ANSI/ISA 1979 ANSI/ISA-S51.1-1979. *Process Instrumentation Terminology*. Instrument Society of America, Research Triangle Park, NC.

Corripio, A.B. 1990. *Tuning of Industrial Control Systems*. Instrument Society of America, Research Triangle Park, NC.

Liptak, B.G. and Venczel, K. 1985a. *Instrument Engineers' Handbook—Process Control*, Revised Edition. Chilton Book Company, Randor, PA.

McMillan, G.K. 1994. *Tuning and Control Loop Performance*, 3rd ed. Instrument Society of America, Research Triangle Park, NC.

National Food Processors Association. 2002. *Validation Guidelines for Automated Control*, Bulletin 43-L. NFPA, Washington, DC

Valentis, K.J., Rotstein, E., and Singh, R.P. 1997. *Handbook of Food Engineering Practice*. CRC Press, LLC, Boca Raton, FL.

Ziegler, J.G. and Nichols, N.B. 1942. Optimum settings for automatic controllers. *Transactions of the ASME*, 64: 759.

## Further reading

ANSI/ISA 1985 ANSI/ISA-S75.01-1985. *Flow Equations for Sizing Control Valves*. Instrument Society of America, Research Triangle Park, NC.

ANSI/ISA 1992 ANSI/ISA-S5.1-1984 (R1992). *Instrument Symbols and Identification*. Instrument Society of America, Research Triangle Park, NC.

ASAE, 1990. Food processing automation. In: *Proceedings of the 1990 Conference, ASAE,* St. Joseph, MI.

ASAE, 1992. Food processing automation. In: *Proceedings of the FPAC II Conference, ASAE,* St. Joseph, MI.

ASAE, 1994. Food processing automation. In: *Proceedings of the FPAC III Conference, ASAE,* St. Joseph, MI.

ASAE, 1995. Food processing automation. In: *Proceedings of the FPAC IV Conference, ASAE,* St. Joseph, MI.

Bauman, H.D. 1994. *Control Valve Primer*, 2nd ed., Instrument Society of America, Research Triangle Park, NC.

Bimbenet, J.J., Dumoulin, E., and Trystram, G. (eds.) 1994. Automatic control of food and biological processes, In: *Proceedings of the ACoFoP III Symposium, Elsevier Science B.V.,* Amsterdam, The Netherlands.

Hughes, T.A. 1988. *Measurement and Control Basics*. Instrument Society of America, Research Triangle Park, NC.

Liptak, B.G. and Venczel, K. 1985. *Instrument Engineers' Handbook—Process Measurement*, Revised Edition. Chilton Book Company, Randor, PA.

Moore Products Co. 1990. *Digital Controller Tuning AM-35*, Issue 3, Moore Products Co., Springhouse, PA.

Morris, R.L. 1982. *Controller Tuning EM-50-14*, Issue 1. Moore Products Co., Springhouse, PA.

Murrill, P.W. 1988. *Application Concepts in Process Control*. Instrument Society of America, Research Triangle Park, NC.

Murrill, P.W. 1991. *Fundamentals of Process Control Theory*, 2nd ed. Instrument Society of America, Research Triangle Park, NC.

St. Clair, D.W. and Freuhauf, P.S. 1994. PID tuning: it's the method, not the rules. *INTECH*, 41(12): 26–30.

# Chapter 3

# PROCESS CONTROL OF RETORTS

*Ray Carroll*

## 3.1. Introduction

There are many types of retorts available to food manufacturers, from the simplest batch, still retort using steam to the various continuous retorts, all having the common purpose of rendering low-acid foods commercially sterile. All retorts, no matter the type, have two fundamental critical control points that must be considered: temperature and time. Depending on the product and the container type and size, a scheduled process, specifying the time that a product must be held at some minimum temperature, is applied to produce foods that are safe for human consumption and free of pathogenic and spoilage organisms capable of growing at normal conditions of storage and distribution.

In this chapter, the control of these two critical factors, as well as the control of all other factors important to the operation of each retort type, will be explored. Process control systems ranging from simple instrumentation to automated control will be considered. In the United States, as well as other countries, there are regulations that stipulate the types of instrumentation a retort must have and the level of control required. It is not the intention here to present government regulations as they apply to retort instrumentation and control, but rather what is necessary for proper control of thermal processes.

*Thermal Processing of Foods: Control and Automation.* Edited by K.P. Sandeep

## 3.2. Critical factors in retort processing

### *3.2.1. Temperature measurement*

The effect of temperature on the destruction of microorganisms can be described by the decimal reduction time, $D_T$. It is measured as the time necessary at a given temperature to reduce the population of the microorganism of concern by 90% or one log. The thermal resistance constant, $Z$, specifies the change in temperature that results in one log change in the Decimal Reduction Time. For example, if a $D_{250}$ value for *Clostridium botulinum* is measured to be 0.2 minute, the population of spores would be reduced by 90% after heating to 250°F for 0.2 minute. If the $Z$ value was determined to be 18°F, then it would take only 0.02 minute to reduce the population of *C. botulinum* spores by 90% at 268°F (Stumbo, 1973). Likewise, a drop in Temperature of 18°F will result in an increase in the time required to reduce the spore population by one log to 2 minutes. This illustrates the critical nature of measuring temperature accurately in a retort. All retorts should be equipped with a reference temperature indicating device and a temperature measuring device used for temperature control. An industry standard for temperature indicating devices is the mercury-in-glass (MIG) thermometer. In the United States, the Food and Drug Administration regulations specify that each retort be equipped with at least one MIG thermometer whose divisions are easily readable to 1°F and whose temperature range does not exceed 17°F per inch of graduated scale (US Department of Health and Human Services, 1996). Outside the United States, other devices, such as resistance temperature detectors (RTD) may be used in lieu of MIG thermometers, when allowed by law. In fact, MIG thermometers may be prohibited due to safety factors arising from glass breakage and mercury contamination, and environmental concerns. All reference temperature indicating devices should be tested for accuracy against a known standard at least once a year.

Because a thermometer measures the temperature at only one point in a retort, the placement of the device in the retort is critical. Temperature indicating devices should be placed close to temperature measuring device used for temperature recording and control. Specific locations will be indicated for each type of retort later in the chapter.

### *3.2.2. Temperature control*

The process temperature within a retort should be controlled by an automated temperature control system and recorded by a temperature/time recording device. The controller and recorder can be combined or be separate devices. The temperature controller provides a means of measuring the process temperature, comparing it to the desired, set-point temperature and adjusting the flow of the heating medium to bring the process temperature to the set point. This may be a simple pneumatic system, or an electronic system controlled by a microprocessor. In the pneumatic system, a temperature bulb inside the retort is directly linked to a bourdon tube via a thermal tube. The bourdon tube will expand when heated and is directly linked to the recorder pen arm and a flapper valve that indirectly adjusts the supply air to open and close the control valve on the heating medium supply line. In the electronic system, a temperature signal is transmitted by an RTD to the controller. An algorithm within the microprocessor compares the temperature to the set point and transmits a signal to adjust the control valve. Control valves are typically proportional valves controlled pneumatically in both pneumatic and electronic systems.

Process cooling may take place completely or partly in the retort. However attained, cooling containers below 110°F is critical to prevent thermophilic spoilage. This temperature should be recorded on the same recording chart, using the same temperature measuring device as for heating in batch retorts.

### *3.2.3. Timing*

As mentioned previously, the process temperature must be maintained for a specified time to provide effective reduction of bacterial spores. Temperature/time recording devices provide a continuous record of process temperature and can provide a means of timing the thermal process. Retort operators should use an analog or digital clock in the process room to verify timing of process steps for batch retorts.

For continuous retorts timing becomes more complex. Time is controlled by the speed of the containers through the retort. How the containers are transported through the continuous retort depends on the type of retort. Control of timing will be achieved by controlling the speed of the motor driving that transport.

### *3.2.4. Pressure*

Each retort must be equipped with a means of measuring pressure within the retort. Correct retort pressure is an indication that the process is correct. Recording the pressure can provide valuable information when troubleshooting a processing deviation. Pressure gauges are also necessary for safety, to indicate abnormally high pressures. Pressure gauges should be graduated in divisions of 2 pounds per square inch or less and positioned so they can be readily observed by the retort operator.

When retorting glass containers, flexible containers such as pouches, or semirigid containers with heat-sealed closures, it is necessary to provide an overpressure—a pressure higher than that required to maintain temperature in the retort. Heat seals are relatively weak, especially at high processing temperatures. Without overpressure control, any fluctuation in steam pressure during processing can upset the balance of internal and external pressure on the container, resulting in excessive stress on the seals. At the onset of cooling in the sterilization cycle, steam inside of the retort collapses, resulting in a rapid drop in pressure. It is critical that overpressure in the retort be accurately controlled at this point and through the cooling cycle to prevent bursting of the heat seals or the package itself. For semi-rigid containers, the internal and external pressures should be balanced as close as possible to within 1 psi to prevent deformation of the container (Figure 3.1). With flexible containers with no defined shape, it is permissible to allow the external pressure to be higher. When processing glass containers, overpressure is necessary not only to prevent breakage, but also to prevent internal pressure from backing the lid off of the container. Too much overpressure must be avoided to prevent the glass finish from cutting through the compound on the lid.

Overpressure should also be applied to cold-packed products sealed without adequate headspace vacuum and processed at high sterilization temperatures (Larousse and Brown, 1997), and cans of

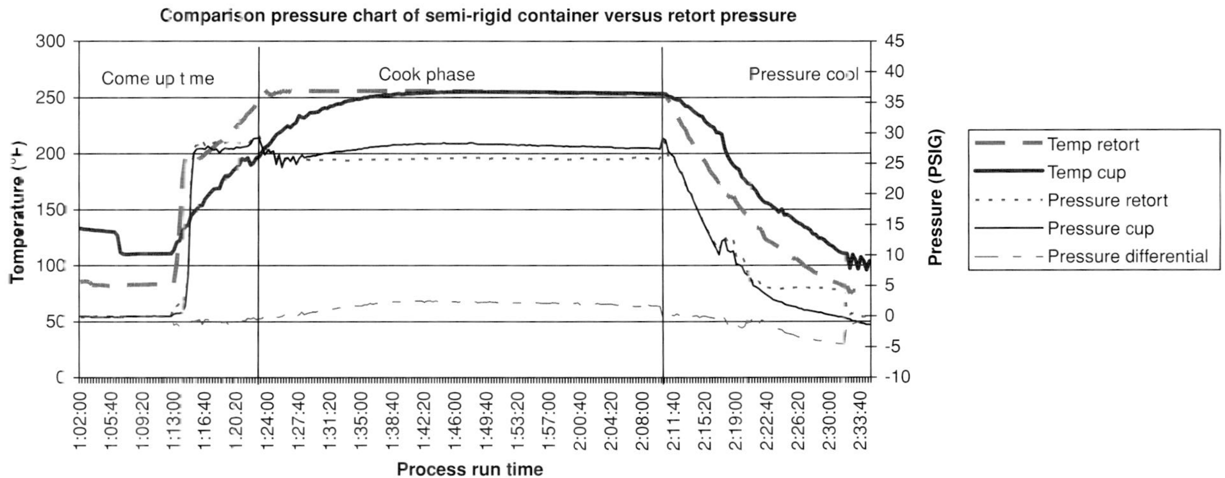


**Figure 3.1.** Pressure differential in a semi-rigid container in a retort.

401 diameter and larger require pressure cooling when processed at 240°F or higher (Gavin and Weddig, 1995). A second, but equally important reason for overpressure during processing of flexible and semi-rigid containers is that gasses within the product can expand and interfere with heat transfer (Ramaswamy and Marcotte, 2006).

Overpressure is achieved through introduction of air or steam into the retort, and must be monitored, recorded, and controlled. As in Temperature control, the recorder and controller can be one unit or two separate devices. The controller can be pneumatic or microprocessor based, or part of a PLC controlling the entire retort process. Overpressure should be maintained within 1 psi of the set point and is controlled through inlet and exhaust valves on the retort.

### *3.2.5. Level*

Retorts that use water immersion as the heating medium must have a means of controlling the level of the water so that all containers are covered with water through the process. It is recommended that the water level be at least 6 inches above the top layer of containers. The retort should be equipped with a sight glass for the operator to verify the water level. In addition, a level sensor equipped with an alarm should be installed.

For hydrostatic retorts, which employ a column of water to maintain temperature and pressure in the processing steam dome, it is critical that the level of water below the steam be maintained within a tight tolerance for proper application of the scheduled thermal process. Improper water level can affect both retort temperature and process time. For this reason, a more sophisticated, automated controller is required. A complete description of hydrostatic retorts and their control systems will be presented later in this chapter.

Several types of retorts, including continuous, agitating retorts require that condensate be removed, or minimum to no level of water be maintained in the vessel for proper operation.

### *3.2.6. Flow rate*

In some types of retorts, it is critical that the heating medium be circulated throughout the vessel to ensure even temperature distribution. This may be achieved via a pump for water or a fan for steam or steam/air mixtures. An indication that the device is operating is required. Pumps should be equipped with a flow switch. For optimal heating, a flow control system utilizing a flow meter is recommended.

### *3.2.7. Temperature distribution testing*

Before applying a thermal process to products in a retort, it is necessary to perform temperature distribution testing in the retort. Temperature distribution tests are performed to ensure that all areas of the retort receive the scheduled process. Beginning with a retort survey, steam supply, number and types of retorts, container size, orientation, and loading configuration are documented. A list, diagram and complete description of the piping, instrumentation, and control systems of the retorts are made (Institute for Thermal Processing Specialists, 1992). Testing is then conducted to determine, among other factors, the venting schedule—the time that vents on a retort must remain open in order to displace air with process steam, the cold spot in the retort, and the come-up time. The placement of vents, drains, temperature measuring devices, and other instrumentation will be addressed for each retort type in the appropriate section. Complete protocols for temperature distribution testing in all types of retorts can be obtained from the Institute for Thermal Processing Specialists.

## 3.3. Classification of retorts

Retorts can be classified in different ways: batch versus continuous, still versus agitating, steam versus superheated water, etc. Batch type retorts will be considered first, and then be subdivided into still versus agitating. The category will be further classified into steam, superheated water, and other considerations such as mixing air with the heating medium.

### *3.3.1. Batch retorts*

In batch retorts, a discreet number of containers are held within the processing vessel for the duration of the thermal process. Therefore,

**Table 3.1.** Process details for an agitating batch retort

| Step | Time Step (minutes) | Elapsed time (minutes) | Temperature (°F) | Pressure (psig) | RPM |
|---|---|---|---|---|---|
| Fill | 2 | 2 | 262 | 0 | 15 |
| Come-up | 9 | 11 | 262 | 28 | 15 |
| Cook | 44 | 55 | 252 | 28 | 15 |
| Pressure cooling | 5 | 60 | 180 | 28 | 15 |
| Atmospheric cooling | 20 | 80 | 75 | 0 | 15 |

temperature is a critical factor and so also the period of time that the temperature is applied to the vessel. Batch retorts tend to be more labor intensive, as containers are usually stacked within baskets or racks, which are then loaded into the retorts. These operations may be manual or completely automated, eliminating the need for human labor. In any case, there is downtime associated with the retort during which the batch is unloaded from the retort and the next batch is loaded. There is also a time period in which the retort is "coming up" to temperature and, likewise, a time period where the retort is cooling down. The periods of loading, come-up, process, cooling, and unloading all make up the cycle time of the retort. Table 3.1 illustrates a typical cycle for an agitating batch retort. It is desirable to minimize the cycle time as much as possible, by optimizing loading, come-up time, and unloading, in order to have greater throughput and more saleable product produced per shift. The process and cooling times, however, must always be controlled as per the scheduled process.

### *3.3.2. Still retorts using steam*

These are the simplest types of retorts available. After the product is loaded and the retort is closed, the steam is turned on. During this time, one or more vents are opened to allow air to be evacuated from the vessel. Vents on vertical and horizontal, still, steam retorts should be installed at the top opposite the steam inlet. Air can act as an insulator, and if not removed from the retort, may cause pockets of low temperature to develop, resulting in underprocessed product. After venting is complete, the vent is closed and a sufficient amount of time is allowed for the retort to reach process temperature. This is known as the come-up time. Venting and come-up time are important process control factors and should be specified in the scheduled process. After completion of the come-up time, the timing of the process begins. With completely manual retorts, the operator must time the venting process, close the vent, observe the temperature, and begin timing the scheduled process after the process retort temperature is met. Temperature should be controlled by an automatic controller as described above. After the prescribed time has passed, the operator will turn off the steam and begin the cooling process. This is achieved by opening the cooling water valve.

When there are multiple batch retorts in an operation, it is common to have a control system that automatically controls the time as well as the temperature of the process and automatically closes the steam valve and opens the cooling water valve. When overpressure is required during cooling, the pressure is automatically controlled and retort temperature is controlled and ramped down gradually. When overpressure is required during processing, it is normally achieved by mixing air with the steam. Automatic pressure control and a fan to maintain a uniform mixture of the steam and air throughout the retort are required. It is essential that the ratio of steam to air be maintained between 75% steam/25% air and 95% steam/5% air in order to provide enough energy to heat the product. The percentage of steam in the retort at any time can be calculated by the following equation:

$$\% \text{ Steam} = \text{steam pressure (psia)/total pressure (psia)}$$

Steam pressure is determined as the pressure of pure saturated steam at the retort temperature.

The temperature indicating device and the temperature transmitter for recording/controlling should be installed through the wall of the retort or in a thermal well attached to the retort, and located close together. Steam enters through a spreader in the bottom of the retort, and a bleeder should be installed at the top to circulate the steam past the temperature devices and allow the operator to observe the steam. Bleeders should be installed in the bottom of the retort to allow for the removal of condensate.

### *3.3.3. Still retorts with water immersion processing*

The operation of these retorts is similar to that of steam retorts. Superheated water is used in place of steam as the processing medium. The level of the water in the retort is a critical factor and must be controlled. The containers must be covered by water throughout the come-up time, processing, and cooling. A level indicating device with an alarm is sufficient. An automatic level control system can prevent low water level process deviations. Vertical retorts employing full water immersion as the processing medium may use compressed air, introduced at the bottom of the retort to circulate the water and ensure proper temperature distribution. For horizontal retorts, it is necessary to use a pump for adequate circulation. Water should be drawn into the suction side of the pump through drains in the bottom of the retort and returned through a spreader that distributes the water evenly through the top of the retort. The temperature indicating device and the temperature measuring device for recording and control should be located together and should be below the water level. The probes should extend at least 2 inches form the wall of the retort into the water. Air or steam may be used to provide overpressure in these retorts when required. In vertical retorts, this may be the same air used for circulation of the water. In horizontal retorts, the air or steam is introduced above the level of the water and is controlled with a pressure sensor, controller, and inlet and exhaust valves.

### *3.3.4. Water spray and water cascading retorts*

Another means of heating containers during processing is through the use of hot water spray systems or alternatively through a water cascading system. In either case, the hot water is recirculated via a pump, and directly contacts the containers. Temperature must be controlled as in other retort types and is achieved through direct injection of steam into the water recirculation line or by an indirect heat exchanger. A minimum water level is maintained at the bottom of the retort below the containers to feed the suction side of the pump. When processing with overpressure, air or steam air mixtures are used to maintain pressure and must be controlled.

The temperature indicating device and the temperature recorder measuring device should be located at the suction side of the recirculating pump, which has been determined to be the cold spot of the water. A separate temperature measuring device used for temperature control may be located at the exit of the heat exchanger. It is very important that temperature distribution testing is done carefully to determine when all parts of the retort are at or above process temperature and timing can commence. Water flow can be measured by a flow meter or alternatively, a pressure differential may be measured between the pressure on the entry and exit ends of the water circulation pump. This pressure differential alarms the control system if the set limits are exceeded. The pressure differential is a measurement of the differences in pressure on the two sides of the pump and may not measure a true water flow. The pressure differential will indicate failure of the water pump but may not always indicate a clogged water distribution system. A flow measuring device is recommended to provide a more accurate measurement of actual water flow in the system. Water flow in the system must be the same as that used during temperature distribution studies in the retort.

### *3.3.5. Agitating batch retorts*

Depending on the consistency of the product and other factors, the rate of heat transfer may be increased by agitating food containers, forcing mixing of the contents, and subsequent convection. This agitation is achieved by a rotating framework inside the retort that holds the baskets, which in turn hold the containers securely during processing. The speed of rotation is a critical factor of the thermal process. It will be specified in the scheduled process and must be controlled. A tachometer is used to measure the rotational speed. It may be used to record the speed and/or control it through the use of a variable frequency drive on the motor. As in still, batch retorts, the heating medium may be water, steam, or steam/air. Air or steam may be used for overpressure when required. The measurement and control of temperature and pressure is required as in still retorts. Locations of temperature measuring and indicating devices are the same. The use and location of vents and bleeders are also the same as for still retorts.

When using full water immersion or water spray/water cascade as the processing medium, an agitating retort requires a pump for recirculation of the water. A means of verifying the operation of the pump—a flow meter, flow switch, or differential pressure measurement is required. Water level must be controlled to cover all containers in full water immersion retorts. Although not a part of the process control of agitating batch retorts themselves, maintaining the proper headspace in the containers is critical to adequate heating and achieving the required lethality.

### 3.3.6. Continuous retort systems

In large thermal processing operations, it is an economic advantage to employ continuous retort systems. Having filled, closed containers transported directly to the retorts via conveyors eliminates the labor (if not automated) and the time required to load batches of containers into and out of the processing vessel. The start-up, venting, and come-up time need to be completed only once at the start of production, rather than for each batch. Manufacturing efficiency is maximized by extending production runs as long as possible. It is also imperative to maintain a constant feed to the continuous retort in order to maximize productivity and to provide sufficient accumulation capacity downstream of the retort; in the case of a stoppage in labeler, case-packer, palletizer, etc., the retort will not have to be shut down.

### 3.3.7. Hydrostatic retorts

For products that heat by conduction, a continuous static retort is employed. The retort must maintain high temperature and pressure ($\sim$250°F and 29.8 psia), while allowing containers to enter and exit. This is achieved by a steam dome balanced by hydrostatic water legs. A pair of parallel chains is connected by carriers that hold the containers and drive them through the retort. Control of the speed of the chain drive is necessary for control of the process time. Speed is controlled by either a mechanical drive or a variable frequency drive on the motor and monitored by a tachometer. The speed should be recorded along with other critical factors on recording charts. It is essential for the operator to manually check the speed of the retort as well and record it on the operator process log. The speed can be checked by counting the carriers that pass by a given point in one minute. The correct speed is determined by dividing the number of carriers in the processing chamber at a given time (specific to the retort) by the process time in minutes. The process time is typically the time that the container is in the steam dome portion of the retort; however, there are instances when lethality credit can be taken in the infeed and/or discharge legs. This will be discussed later.

Many of the other critical factors in hydrostatic retort processing are interrelated, making the control system more complex than in batch retorts. There is no opportunity to provide overpressure during processing; however, control of system pressure is necessary to maintain temperature. Maintenance of pressure is controlled not only by the steam control valve but also by the levels of the hydrostatic water legs. The height of the legs and subsequently the level of water under the steam dome will influence the residence time in the dome. Steam constantly condenses on the containers, adding to the water level; at the end of the discharge leg, containers carry a small amount of water out of the system. As containers exit the hydrostatic retort, an equal number of containers enter the retort simultaneously. This helps to maintain a constant level. When the infeed to the hydrostatic retort is starved, the balance is upset; the displacement of the containers exiting the system is eliminated because there are no new containers to replace them in the system. Consequently, the water level drops. Water must be added to maintain the level. Similarly, when the feed to the retort is reinstated, and the void reaches the discharge leg, an equal amount of water must then be removed from the system.

#### *3.3.7.1. Level control*

Water level is controlled at the base of the hydrostatic retort. The area where the steam in the dome contacts the water is called the steam–water interface. It is the height of this interface that must be controlled primarily to ensure proper control of the process time. Should the water level rise, the process time is decreased. Should it fall, the process time is lengthened. It is also important to ensure that containers do not contact the water when the chain turns at the bottom of the steam dome to begin climbing back to the top. Should this occur, the temperature of the product may drop, resulting in a process deviation. A level sensor should be located at the interface that signals control valves to add or release water from the system. A high-level alarm should be provided at the lowest point that the container carriers reach, signaling the operator that a deviation has occurred. A sight glass should be provided for the operator to observe and record the water level.

*3.3.7.2. Temperature control*

Because of the complexity mentioned above, the control of temperature in hydrostatic retorts must be automated. As in other retorts, a reference thermometer should be provided as well as a temperature measuring device used for recording and controlling the temperature. Both of these devices should be located at the bottom of the steam dome near the steam–water interface. This is the coldest spot of the processing chamber. The automated system should operate a steam control valve to maintain process temperature.

If lethality credit is taken for the feed or discharge legs, the water temperature must be monitored and controlled in these legs as well. A reference thermometer should be provided as well as a temperature measuring device used for recording and controlling the temperature of the leg. Temperature is controlled by injecting steam into the water at the base of the leg. It is common practice to monitor and record the temperature at several locations of the processing chamber and each leg. A typical recording chart station is shown in Figure 3.2.

Container cooling begins in the discharge leg and continues in an atmospheric cooling leg utilizing a water spray system. It is imperative that containers are cooled below 110°F after processing to prevent the growth of thermophilic spoilage organisms.

The pressure will be controlled automatically by the temperature and the height of the hydrostatic water legs. A pressure indicating device should be located in the steam dome. A pressure recorder is optional, but recommended for troubleshooting and diagnosing process deviations.

At the beginning of production, the hydrostatic retort must be vented, like other retorts, to remove air from the processing chamber. The vent should be adjacent to the entry of the steam. Also, similar to other steam retorts, the hydrostatic retort should be equipped with

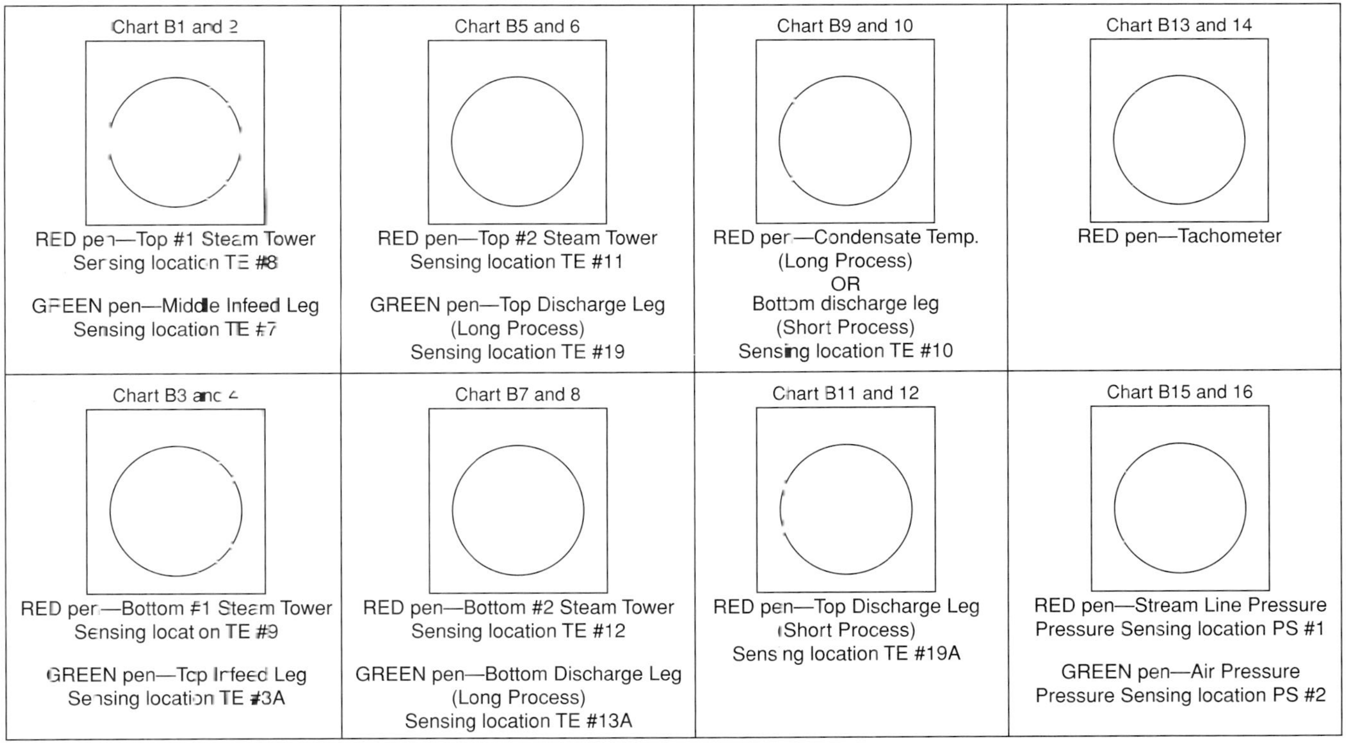


**Figure 3.2.** Hydrostatic retort control charts.

bleeders to allow a small stream of steam to exit the retort. Bleeders should be located adjacent to the entry of steam and should be visible to the operator.

### *3.3.8. Continuous agitating retorts*

Products that can be heated more rapidly by forced convection and are packaged in cylindrical containers (metal cans) can be processed most efficiently with continuous agitating retorts (Fig. 3.4) employing steam as the heating medium. As in hydrostatic retorts, it is necessary to maintain the temperature and pressure in the processing chamber while allowing containers to enter and exit. Rather than passing the containers through water columns that trap the steam inside a dome, the continuous agitating retorts are closed, cylindrical, pressure vessels equipped with valves that transfer containers into and out of the retort (Figure 3.3). The valves act as pressure locks between the atmospheric pressure of the room and the pressurized retort shell. The retort uses a rotating reel that contains evenly spaced steps to hold containers against the inner wall of the vessel or shell. A spiral tee is used to advance the containers through the retort, one container height each revolution. The speed of the rotating reel dictates the

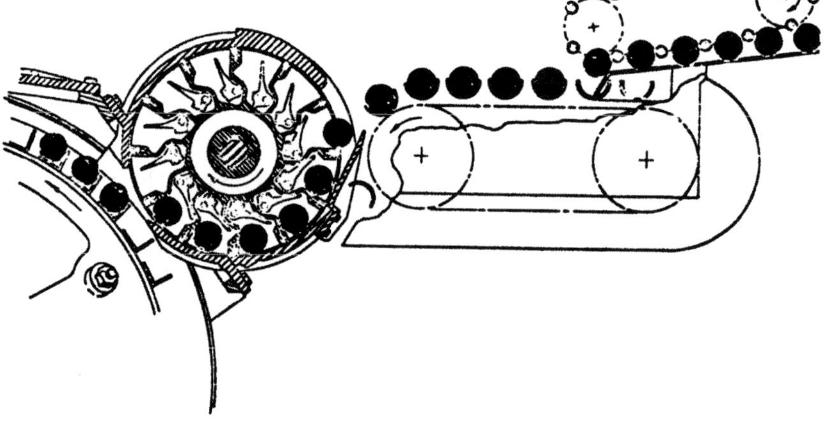

**Figure 3.3.** Container entry transfer valve (FMC Corp.).

**Figure 3.4.** Continuous agitating retort (FMC Corp.).

process time. Speed is controlled by a variable frequency drive on the motor. When the containers are in the upper section of the vessel during each revolution, they are carried by the steps. When they are in the lower section, they roll freely on the inner wall of the vessel. It is during this rolling period that the agitation occurs. It is necessary to maintain a headspace in the closed container to allow the bubble to move through and mix the food product. Rotational speed effects not only the process time, but also the degree of agitation. It should be checked manually once per shift by counting rotations in 1 minute, and recorded on the operator log.

*3.3.8.1. Temperature control*

Temperature is controlled by an automatic control system that employs a temperature measuring device and a steam control valve. Similar to other retort systems, a separate temperature indicating device should be installed next to the temperature measuring device used for control, for operator verification. Process temperature should be recorded on an appropriate recording chart. The temperature sensors should be located together with the bulbs installed in the pressure vessel or an external well. Steam is introduced into the retort through a manifold at the bottom of the vessel. Condensate must be removed from the vessel to prevent interference with container rotation. It is common for a continuous agitating retort to have multiple shells to provide longer process times. Each shell should have its own temperature measurement and control system, and should follow the guidelines outlined above.

Cooling of the containers is conducted in separate shells that operate in the same manner as the heating shells, using cold water in place of steam. When necessary, overpressure can be provided in the cooling section by introducing compressed air into the vessel. This pressure must be controlled and maintained at a pressure lower than the heating vessel to prevent air from leaking into the heating vessel through the transfer valve. A pressure sensor in the wall of the vessel is used to control the air inlet valves. It is common to have a pressure cooling shell followed by an atmospheric cooling shell.

At the beginning of production, each heating vessel must be vented to remove air. During processing, the operator should observe both steam operation and condensate removal via bleeders installed through the vessel walls.

## References


Gavin, A. and Weddig, L. 1995. *Canned Foods, Principles of Thermal Process Control, Acidification and Container Closure Evaluation*, 6th ed. The Food Processors Institute, New York.

Institute for Thermal Processing Specialists. 1992. Available at: http://iftps.org/protocols.html.

Larousse, J. and Brown, B. 1997. *Food Canning Technology*, Wiley-VCH, New York.

National Canners Association. 1968. Laboratory Manual for Food Canners and Processors. AVI, Westport, CT.

Ramaswamy, H. and Marcotte, M. 2006. Food Processing Principles and Applications. CRC Press, Boca Raton, FL.

Scott, G. 1992. The Application of Heat Penetration Models to the Dynamic Simulation of a Hydrostatic Retort, Technical Memorandum No. 657, Campden Food and Drink Research Association, Chipping Campden, UK.

Stumbo, C.R. 1973. *Thermobacteriology in Food Processing*, 2nd ed. Academic Press, New York.

US Department of Health and Human Services, Food and Drug Administration, 1996. Requirements for Establishment Registration, Thermal Process Filing and Good Manufacturing Practice for Low Acid Canned Foods and Acidified Foods.

# Chapter 4

# ON-LINE CONTROL STRATEGIES TO CORRECT DEVIANT THERMAL PROCESSES: BATCH STERILIZATION OF LOW-ACID FOODS

*Ricardo Simpson, I. Figueroa, and Arthur A. Teixeira*

## 4.1. Introduction

Thermal processing is an important method of food preservation in the manufacture of shelf stable canned foods, and has been the cornerstone of the food processing industry for more than a century. Thermal process calculations, in which process times at specified retort temperatures are calculated in order to achieve safe levels of microbial inactivation (lethality), must be carried out carefully to assure public health safety (Bigelow *et al.*, 1920; Ball, 1928; Stumbo, 1973; Pham, 1987; Teixeira, 1992; Holdsworth, 1997). However, overprocessing must be avoided because thermal processes also have a detrimental effect on the quality (nutritional and sensorial factors) of foods. Therefore, the accuracy of the methods used for this purpose is of importance to food science and engineering professionals working in this field. Control of thermal process operations in food canning factories has consisted of maintaining specified operating conditions that have been predetermined from product and process

*Thermal Processing of Foods: Control and Automation.* Edited by K.P. Sandeep

heat penetration tests, such as the process calculations for the time and temperature of a batch cook. Sometimes unexpected changes can occur during the course of the process operation such that the prespecified processing conditions are no longer valid or appropriate; and off-specification product is produced that must be either reprocessed or destroyed at appreciable economic loss. These types of situations are known as process deviations. Because of the important emphasis placed on the public safety of canned foods, processors must operate in strict compliance with the US Food and Drug Administration's Low-Acid Canned Food (FDA/LACF) regulations. Among other things, these regulations require strict documentation and record keeping of all critical control points in the processing of each retort load or batch of canned product. Particular emphasis is placed on product batches that experience an unscheduled process deviation, such as when a drop in retort temperature occurs during the course of the process, which may result from loss of steam pressure. In such a case, the product will not have received the established scheduled process, and must be either fully reprocessed, destroyed, or set aside for evaluation by a competent processing authority. If the product is judged to be safe, then batch records must contain documentation showing how that judgment was reached. If judged unsafe, then the product must be fully reprocessed or destroyed. Such practices are costly.

According to the US Food and Drug Administration (FDA), a retort temperature deviation occurs when the retort temperature drops more than 0.5°C below the one specified in the process registration file (Larkin, 2002). Therefore, processors of low-acid canned foods must have an effective and efficient control system over the retort sterilization process to avoid unexpected process deviations that would leave the resulting process lethality in question. With this kind of system, smooth uninterrupted operation of a retort battery system in large cook rooms is assured and human error due to manual operation is reduced. In general, commercial on-line control systems consist of three parts: data acquisition, computer software with control and computational algorithms, and the interface hardware. Only one computer is required to control a large number of retorts. The real-time display is available to the supervisor in graphic mode, may also be permanently stored in user-defined files, and can provide both automatic and manual control.

### 4.1.1. On-line control strategies

Commercial systems currently in use for on-line retort control accomplish on-line correction of process deviations by extending process time to that which would be needed had the entire process been carried out at the retort temperature reached at the lowest point in the deviation. These alternative retort temperature–time combinations are predetermined for each product from heat penetration tests, and stored in a file for immediate access when needed. This method of correction often results in significant unnecessary overprocessing with concomitant deterioration in product quality. Further, the extended process time that is given can cause a costly interruption to the retort loading/unloading rotation schedule in large cook room operations, with negative impact on manufacturing efficiency.

Nonetheless, these commercial systems are versatile because they are applicable to any kind of food under any size, type, or shape of container, as well as mode of heat transfer (Larkin, 2002). They also require no mathematical heat transfer models for predicting cold spot temperature in response to variable retort temperature. In addition to the versatility and relative simplicity of this control strategy, it is also clear that it will always result in a safe correction, but by no means optimal or efficient (Alonso *et al.*, 1993; Von Oetinger, 1997; Simpson, 2004).

### 4.1.2. On-line control strategies applicable to mathematically modeled low-acid foods

Recognizing this limitation, a control logic algorithm, first conceived by Teixeira and Manson (1982) and further developed by Datta *et al.* (1986), was proposed for use with computer-based control systems on batch retort operations, that incorporated a mathematical heat transfer model to simulate heat conduction into a solid body in the shape of a finite cylinder. The proposed system was capable of automatically adjusting process time during the cook cycle to compensate for any unexpected deviation in retort temperature. The model was used to predict the product cold spot temperature in response to input readings of the measured retort temperature. As the predicted cold spot temperature profile developed over time, the accumulating lethality was calculated by the general method in real time, and compared with the target value required at the end of heating. Once the calculated lethality agreed with the target value, the control signal to end heating and begin cooling was given. By programming the control logic to continue heating until the accumulated lethality had reached some designated target value, the process would always end with the desired level of final lethality regardless of any unscheduled process temperature deviation. The advantage of this system was that it was capable of implementing an exact or required correction for a process deviation without unnecessary overprocessing. However, the mathematical heat transfer model proposed for this system was only applicable to conduction-heated foods packed in cylindrical containers. These limitations were later overcome by replacing the finite cylinder conduction model with an alternative model that applied to any shape container or mode of heat transfer (Noronha *et al.*, 1995; Teixeira *et al.*, 1999). In either case, although these models produced an exact or precise correction (no overprocessing), the final process time would always be longer than the original process time, and could still cause work schedule interruptions in large cook room operations with batteries of retorts on scheduled process rotation cycles. Recently published reviews by Teixeira and Tucker (1997), as well as Simpson *et al.* (2004), have pointed to the need for new and novel on-line retort control strategies that can address this and other issues.

The first part of this chapter presents an optimum on-line correction of process deviations in batch retorts through simulation complying with the following:

- Exact on-line correction of an unscheduled process deviation without extending the originally scheduled process time to maintain the synchronous scheduling retort cycle times within the battery of retorts in a large cook room operation.
- Exact on-line correction of process deviations occurring during a preprogrammed variable or dynamic retort temperature process,

such as a ramp-up and ramp-down profile, that might be chosen to maximize nutrient retention in a conduction-heated food; and to do so with minimum compromise to the required level of final nutrient retention.

### *4.1.3. On-line control strategies applicable to any low-acid foods*

Based on observations and record review of various retort control systems, the FDA concluded that the design of alternative-schedule retort control systems (table-lookup method) meet the overall intent of the FDA/ LACF regulations at 21 CFR 113, and appear to provide safeguards that are sufficiently equivalent to the safeguards provided by the mandatory provisions of these regulations. Table-lookup methods are easily implemented in any cannery plant, and are a safe FDA-approved on-line procedure for correct process deviations. However, they are a very inefficient way to correct a deviation because they often result in extensive overprocessing resulting in poor product quality and costly disruption of cook room retort operating schedules. In order to avoid these inefficient corrections, processors tend to operate 1°C or 2°C, and some times 3–4°C, over the registered retort temperature. Therefore, processors are able to minimize, even reduce to zero the occurrence of deviant processes but with the cost of exposing every single batch to an overprocessing situation.

In the second part of this chapter, a safe, simple, efficient, and easy-to-use procedure is presented to manage on-line corrections of unexpected process deviations in any canning plant facility complying with the following:

1. Strategy able to correct the process deviation by an alternative "proportional- corrected" process that delivers no less than final target lethality, but with near minimum extended process time at the recovered retort temperature.
2. Demonstrate strategy performance by comparing "proportional-corrected" with "commercial-corrected" and "exact-corrected" process times.
3. Demonstrate consistent safety of the strategy by exhaustive search over an extensive domain of product and process conditions to find a case in which safety is compromised.

## 4.2. On-line control strategies for low-acid foods that can be mathematically modeled

### *4.2.1. Correction of a deviant process while maintaining the preestablished process time*

The objective for the control strategy required in this task (Task 4.2.1) was to accomplish an on-line correction of an unexpected retort temperature deviation without changing the originally scheduled retort process time. Fundamental to all strategies presented in this paper is the understanding that the retort control system would include a computer that is running the software containing the appropriate mathematical heat transfer model, and that the computer continually reads the actual retort temperature from a temperature-sensing probe through an analog/digital (A/D) data acquisition system. This continual reading of retort temperature would be used as real-time input of dynamic boundary condition for the mathematical heat transfer model. The model, in turn, would be accurately predicting the internal product cold spot temperature profile as it develops in response to the actual dynamic boundary condition (retort temperature), just as described earlier by Datta *et al.* (1986). As the predicted cold spot temperature profile develops over time, the accumulating lethality ($F_0$) would be calculated by the general method and known at any time during the process. Should a deviation occur during the process, a simulated search routine would be carried out on the computer to find the combination of process conditions for the remainder of the process that would result in meeting the final target lethality without overextending the process time.

The key in this strategy was to identify the retort temperature as the control variable to be manipulated during the remainder of the process (rather than process time). Therefore, upon recovery of the deviation, the search routine would find the new higher retort temperature to be used for the remainder of the process, and send the appropriate signals through the data acquisition system to readjust the retort temperature accordingly. This entire procedure would be triggered when retort temperature would fall more than 0.5°C below the one specified in the FDA process registration file.

Increasing retort temperature cannot be accomplished without increasing steam pressure correspondingly, which dictates a practical upper limit to choice of higher retort temperature. An upper limit of 135°C was chosen for the simulations carried out in this work. This upper limit comes into play when the deviation occurs near the end of the process, when the little time remaining forces the simulation search routine to choose the upper limit for retort temperature. In these cases, the safety requirement for reaching the final target lethality ($F_0$) must take priority over compromising process time. This will inevitably require some extension in process time, but it will be an absolute minimum, that would not likely upset scheduling routines. Examples of this control strategy being applied for correcting deviations in the process (both early and late), were simulated for the case of thermal processes typical of still-cook conduction heating in cylindrical can or retort pouch, and forced convection heating in cylindrical can under mechanical agitation.

*4.2.1.1. Simulation conditions for Task 4.2.1*

For Task 4.2.1 three simulation cases were demonstrated:

1. Still-cook conduction heating of solid product in cylindrical can.
2. Still-cook conduction heating of solid product in retort pouch.
3. Forced-convection heating of liquid product in cylindrical can under mechanical agitation.

Each case was simulated twice. The first simulation is with the retort temperature deviation occurring near the midpoint of the process, and the second with the deviation occurring near the end of the process, which would be the most challenging situation. The product and process conditions chosen to carry out the demonstrated simulations for each case in Task 4.2.1 are given in Table 4.1.

Three simulation cases carried out for Task 4.2.1 are shown graphically in Figures 4.1–4.3. Each figure shows the profile of the original scheduled retort temperature and the profile of the simulated deviation, as well as the new corrected retort temperature profile. In addition, the accumulated lethality calculated over time is also given, with the scale for lethality presented at right-hand side of the graph. Figures 4.1a and 4.1b are for the case of conduction-heating solid

**Table 4.1.** Product and process conditions used for control strategy simulations in Task 4.3.1 (on-line correction of deviations without change in process time)

| | Container dimensions (cm) | | | Properties | | Normal process | |
|---|---|---|---|---|---|---|---|
| Product simulated | Major | Intermediate | Minor | Alpha ($m^2/s$) | $f_h$ (minutes) | Time (minutes) | TRT (°C) |
| Pure conduction can, Biot number > 40 | 11.3 | — | 7.3 | $1.70\times10^{-7}$ | 44.4 | 64.1 | 120 |
| Pure conduction pouch, Biot number > 40 | 15 | 12 | 3 | $1.70\times10^{-7}$ | 20.6 | 35.4 | 120 |
| Forced convection can, Biot number < 1 | 11.3 | — | 7.3 | — | 4.4 | 15.6 | 120 |

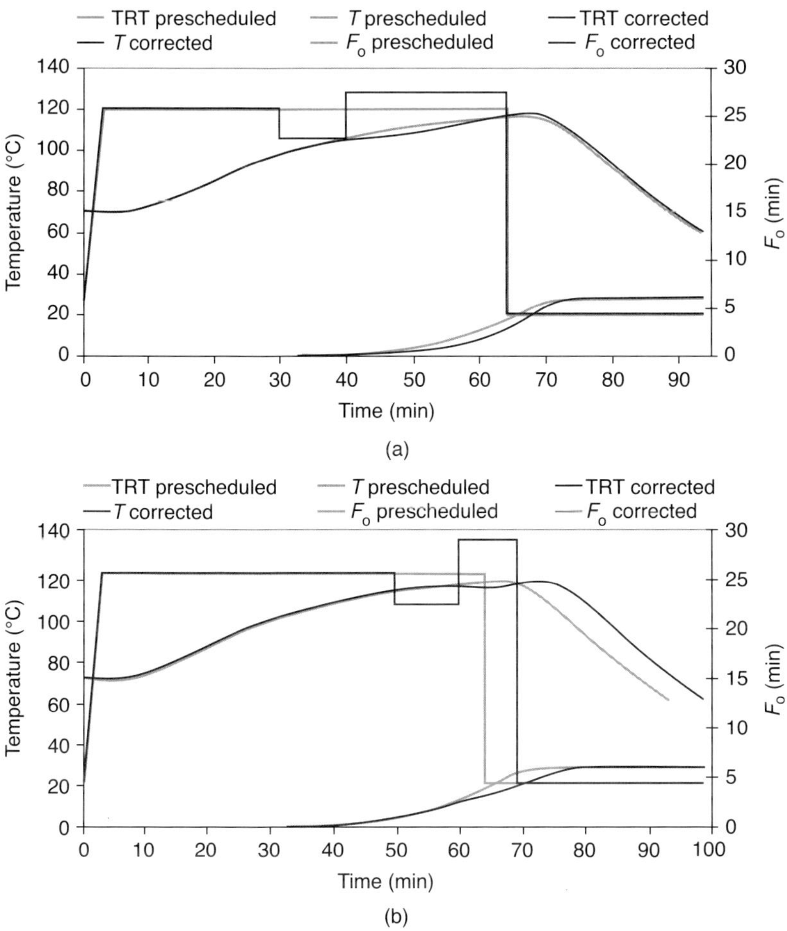


**Figure 4.1.** (a) Pure conduction simulation for on-line correction of an unexpected retort temperature deviation occurring approximately *midway* into the scheduled process for a cylindrical can of conduction-heated food. (b) Pure conduction simulation for on-line correction of an unexpected retort temperature deviation occurring near the *endpoint* of the scheduled process for a cylindrical can of conduction-heated food.

food in a cylindrical can. Figure 4.1a shows the result for on-line correction of an unexpected retort temperature deviation occurring after 30 minutes, or approximately midway into the scheduled process. Note that the normal process was originally scheduled for 64 minutes at retort temperature of 120°C. The deviation lasted for duration of 10 minutes and reached a low point of 105°C. Yet, the correction was accomplished within the originally scheduled 64 minutes by completing the duration of the process at a new higher retort temperature of 127°C. Figure 4.1b shows the case of when the same extent of deviation occurs near the end of the process when insufficient time remains to reach the final target lethality within the upper limit of retort temperature that is possible (135°C). In this case, process time must be extended in order to achieve the safe final target lethality, but it will be an absolute minimum without compromising safety. Likewise, Figures 4.2a and 4.2b and 4.3a and 4.3b show similar results for the case of forced-convection heating of liquid product in cylindrical can under mechanical agitation, and still-cook conduction heating of solid product in retort pouch, respectively.

### *4.2.2. Correction of a deviant process (variable retort temperature process) with a variable temperature profile while maintaining process time and maximizing a quality factor*

The objective for the control strategy required in this task (Task 4.2.2) was to accomplish an on-line correction of an unexpected retort temperature deviation while maximizing the final quality retention in the product, and to demonstrate the application of this strategy to dynamic (time varying) retort temperature processes. This strategy is meant to apply to those retort processes that were designed specifically for retaining maximum quality, such as nutrient retention, in the product. Canned foods in this category are usually slow conduction-heating products in relatively large containers in which significant nonuniform temperature distributions exist throughout the thermal retort process. These nonuniform temperature distributions provide opportunity to improve levels of quality retention by use of dynamic or time-variable retort temperature processes over traditional constant retort temperature processes. Results from process optimization studies have shown that most optimum retort temperature profiles for this purpose involve a ramp increase to a maximum retort temperature followed by a ramp decrease to a minimum prior to cooling (Teixeira *et al.*, 1975; Saguy and Karel, 1979). For these types of products the heat transfer model must also be able to accurately calculate the quality degradation caused by the thermal process. The

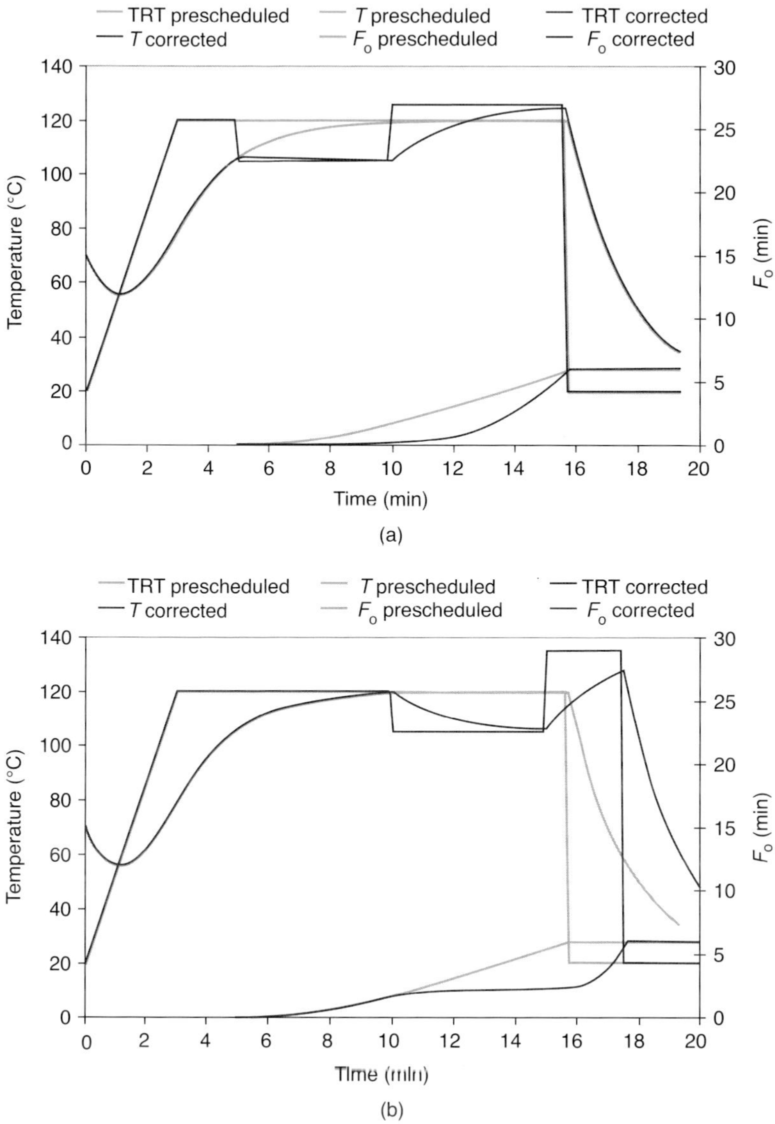


**Figure 4.2.** (a) Forced convection simulation for on-line correction of an unexpected retort temperature deviation occurring approximately *midway* into the scheduled process for a cylindrical can of liquid product under forced convection by mechanical agitation (perfect mixing). (4b) Forced convection simulation for on-line correction of an unexpected retort temperature deviation occurring near the *endpoint* of the scheduled process for a cylindrical can of liquid product under forced convection by mechanical agitation (perfect mixing).

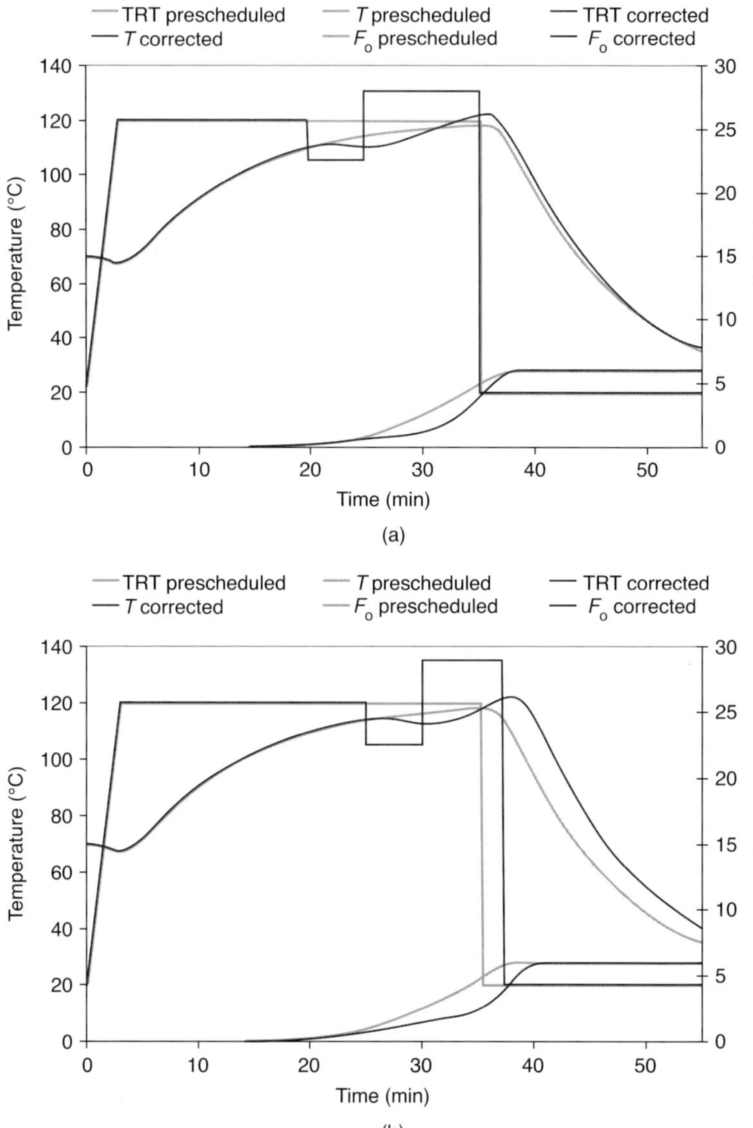


**Figure 4.3.** (a) Retort pouch simulation for on-line correction of an unexpected retort temperature deviation occurring approximately *midway* into the scheduled process for a solid conduction-heated food in shape of thin slab. (b) Retort pouch simulation for on-line correction of an unexpected retort temperature deviation occurring near the *endpoint* of the scheduled process for a retort pouch of solid conduction-heated food in shape of thin slab.

thermal degradation kinetics of most quality factors are such that these calculations are very sensitive to the nonuniform temperature distribution in such conduction-heated foods. Therefore, the conduction heat transfer models to be used for this purpose must be shape specific for the shape of product container to be simulated. This not necessarily the case when only lethality at the cold spot is of concern, and will be discussed further in connection with construction of heat transfer models.

The focus at this point is on use of the model's ability to calculate volume-integrated mass-average nutrient retention (as described by Teixeira *et al.*, 1969) in response to time-varying boundary conditions (variable retort temperature). In the case of an unexpected retort temperature deviation occurring during the preprogrammed ramp-up or ramp-down of a variable retort temperature process the strategy would call for completing the process after recovery of the deviation at the optimum constant retort temperature process time combination that would maximize final nutrient retention. Just as in the previous case in Task 4.2.1, the deviation would trigger a simulation search routine of finding a family of remaining retort temperature–time combinations that would deliver the same final target lethality, and choose the one that yielded maximum nutrient retention. Examples of this control strategy being applied for correcting deviations in the process (during ramp-up and ramp-down) were simulated for the case of a thermal processes typical of a still-cook conduction-heating product in a cylindrical can (same as that used in Task 4.2.1).

#### *4.2.2.1. Simulation conditions for Task 4.2.2*

Task 4.2.2 only applies to a conduction-heating solid product in which a dynamic retort temperature process might be specified as the normal process to maximize the final quality retention in the product. In this case, cylindrical container of the same size and shape as that used for the conduction-heating finite cylinder model in Task 4.2.1 was also used in Task 4.2.2, as well as assuming a solid product with the same physical and thermal properties (alpha and $f_h$). These data, along with the specific process conditions of normal and corrected process times and nutrient retentions, are given in Table 4.2.

**Table 4.2.** Product and process conditions used for control strategy simulations demonstrated in Task 4.3.2 (on-line correction of variable retort temperature process without compromising quality)

| | Container dimensions | | | Properties | | |
|---|---|---|---|---|---|---|
| Product simulated | Major (cm) | Intermediate (cm) | Minor (cm) | Alpha ($m^2/s$) | $f_h$ (minutes) | Nutrient kinetics |
| Pure conduction can, Biot number > 40 | 11.3 | – | 7.3 | $1.70 \times 10^{-7}$ | 44.4 | $D_{121} = 250$ minutes; $z = 25°C$ |

| | Time (minutes) | | $F_o$ (min) | | Nutrient retention | |
|---|---|---|---|---|---|---|
| Case simulated | Normal | Corrected | Normal | Corrected | Normal | Corrected |
| Early deviation (15 minutes) | 51.3 | 60.2 | 6.04 | 6.04 | 0.704 | 0.702 |
| Late deviation (35 minutes) | 51.3 | 61.2 | 6.04 | 6 | 0.704 | 0.705 |

The simulation cases that were demonstrated for Task 4.2.2 are shown graphically in Figures 4.4a and 4.4b. Recall that Task 4.2.2 only applies to a conduction-heating solid product in which a dynamic retort temperature process might be specified as the normal process to maximize the final quality retention in the product. In this case, cylindrical container of the same size and shape as that used for the conduction-heating finite cylinder model in Task 4.2.1 was also used in Task 4.2.2, as well as assuming a solid product with the same physical and thermal properties (alpha and $f_h$). Figure 4.4a shows the result for on-line correction of an unexpected retort temperature deviation occurring midway during the ramp-up portion into the scheduled process, while Figure 4.4b shows the result for a deviation occurring during the ramp-down portion. Note that in both cases process time had to be extended about 10 minutes in order to deliver the same required final lethality without any significant compromise in quality (nutrient) retention. This is the price that must be paid to avoid compromise in quality. Note that the original nutrient retention expected from the normal scheduled process as well as that finally delivered by the longer corrected processes were shown earlier in Table 4.2, and are essentially the same.

## 4.3. On-line control strategy for any low-acid food

The aim of this strategy is to attain a safe, simple, efficient, and easy to use procedure to manage on-line corrections of unexpected process deviations in any canning plant facility. Specific objectives are:

- Developing strategy to correct the process deviation by an alternative "proportional- corrected" process that delivers no less than final target lethality, but with near minimum extended process time at the recovered retort temperature.
- Demonstrating strategy performance by comparing "proportional-corrected" with "commercial-corrected" and "exact-corrected" process times.
- Demonstrating consistent safety of the strategy by exhaustive search over an extensive domain of product and process conditions to find a case in which safety is compromised.

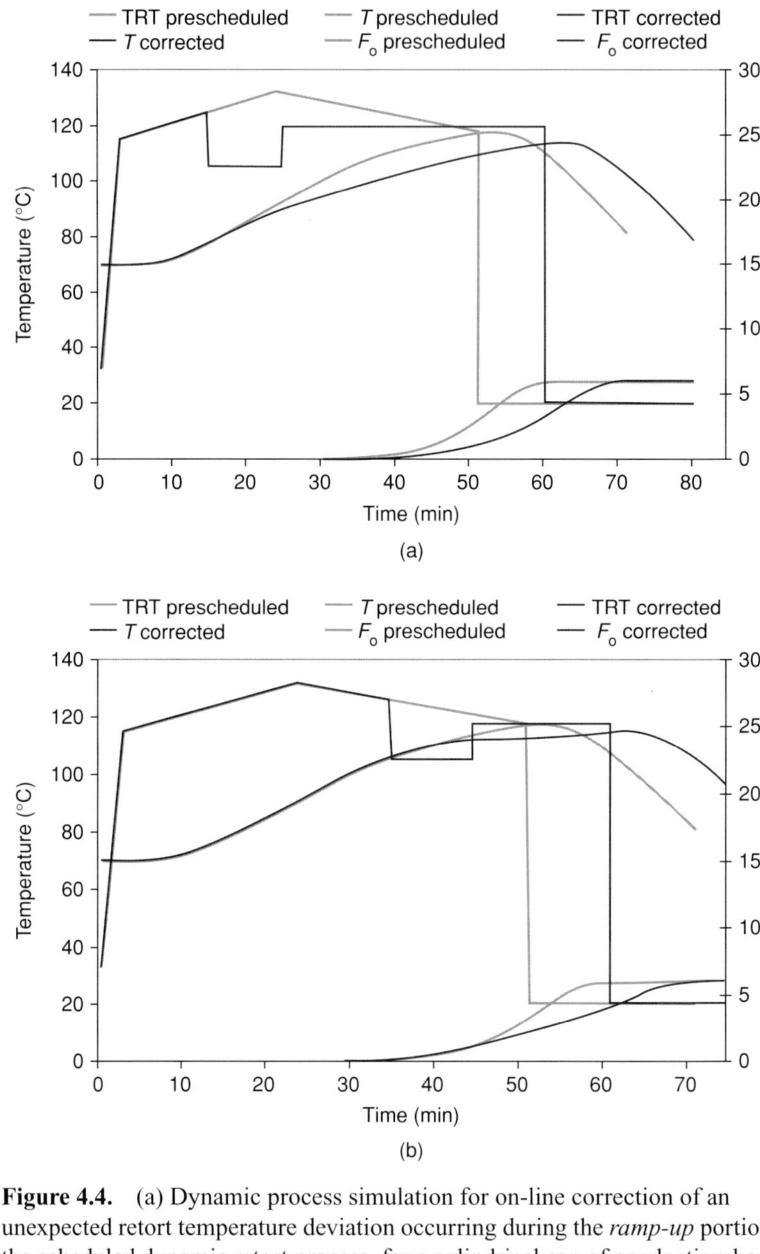


**Figure 4.4.** (a) Dynamic process simulation for on-line correction of an unexpected retort temperature deviation occurring during the *ramp-up* portion into the scheduled dynamic retort process, for a cylindrical can of conduction-heated food. (b) Dynamic process simulation for on-line correction of an unexpected retort temperature deviation occurring during the *ramp-down* portion into the scheduled dynamic retort process, for a cylindrical can of conduction-heated food.

To attain the objectives stated above, the approach to this work was carried out in four tasks, one in support of each objective. Task 4.3.1 consisted of developing the strategy for on-line correction of process deviations with minimum extended process time using the method of "proportional correction." Task 4.3.2 consisted of choosing appropriate mathematical heat transfer models for construction of the equivalent lethality curves or "look-up tables" needed for use with each respective strategy, and for determining the final lethality and quality retention for each of the thousands of cases simulated in the study. Task 4.3.3 included the complex optimization search routine that was carried out to demonstrate validity and consistent safety of the strategy. Methodology employed in carrying out each of these tasks is described in greater detail below.

### *4.3.1. Proportional-correction strategy development*

The objective for the strategy required in this task was to accomplish an on-line correction of an unexpected retort temperature deviation by an alternative process that delivers final target lethality, but with minimum extended process time at the recovered retort temperature. This would be accomplished with use of the same alternative process "look-up tables" that would normally be used with currently accepted methods of on-line correction of process deviations, but with a "proportional correction" applied to the alternative process time that would reduce it to a minimum without compromising safety. In order to fully understand this strategy, it will be helpful to first review the currently accepted method that is in common practice throughout the industry. Commercial systems currently in use for on-line correction of process deviations do so by extending process time to that which would be needed to deliver the same final lethality had the entire process been carried out with an alternative lower constant retort temperature equal to that reached at the lowest point in the deviation. These alternative retort temperature–time combinations that deliver the same final process lethality ($F_0$) are called equivalent lethality processes. When these equivalent time–temperature combinations are plotted on a graph of process time versus retort temperature, they fall along a smooth curve called an equivalent lethality curve. These curves are predetermined for each product from heat penetration tests and thermal process calculations carried out for different retort temperatures, and stored in a file or on a chart for immediate access when needed. Many food canning plants maintain updated files with such equivalent lethality curves prepared for each scheduled process for every product manufactured.

In practice, the new process times obtained from these curves at such low alternative temperatures can be as much as two or three times longer than the originally scheduled process time required to reach the same final target lethality, resulting in considerable quality deterioration and costly disruption to scheduled retort operations. These consequences are particularly painful when, as in most cases, the deviation recovers quickly, and the alternative extended process time is carried out at the recovered original retort temperature. Canned food products subjected to such "corrected" processes become severely overprocessed, with final lethalities far in excess of that required and quality deterioration often reaching levels below consumer acceptance. To avoid these painful corrections, processors normally operate at retort temperatures 3–4°C over the registered retort temperature.

The "proportional-correction" strategy developed in this paper significantly avoids such excessive overprocessing by taking advantage of the short duration of most recovered retort temperature deviations, and the lethality delivered by carrying out the corrected process at the recovered retort temperature. The strategy will calculate the corrected process time ($t_{\mathrm{D}}$) as a function of the temperature drop experienced during the deviation, and also the time duration of the deviation. The following expression illustrates mathematically how this "proportional-corrected" process time would be calculated for any number ($n$) of deviations occurring throughout the course of a single process:

$$t_{\mathrm{D}} = t_{\mathrm{TRT}} + \sum_{i=1}^{n} (t_{\mathrm{D}_i} - t_{\mathrm{TRT}}) \left( \frac{\Delta t_i}{t_{\mathrm{TRT}}} \right) \qquad t_{\mathrm{D}_i} \geq t_{\mathrm{TRT}}$$

where $n$ is the number of deviations occurring during the process, $t_{\mathrm{D}}$ is corrected process time, $t_{\mathrm{TRT}}$ is preestablished process time at retort temperature TRT, $\Delta t_i$ is the duration of deviation $i$, $t_{\mathrm{D}i}$ is the process time at the deviation temperature $TRT_i$, $TRT_i$ is the lowest temperature during the deviation $i$, and TRT is the retort temperature.

For example, in the case of a single deviation, the corrected process time would be calculated by first finding the alternative process time that would be required to deliver the same final lethality had the entire process been carried out with an alternative lower constant retort temperature equal to that reached at the lowest point in the deviation ($t_D$). This would be done by use of the equivalent lethality process curve or look-up table described earlier. The difference between this longer alternative process time and the originally scheduled process time ($t_D - t_{TRT}$) is the extra time that would normally be added to the original process time to correct the process according to current industry practice. However, in the new strategy proposed here, this extra process time differential ($t_D - t_{TRT}$) is multiplied by a proportionality factor consisting of the ratio, [time duration of the deviation]:[originally scheduled process time], or expressed mathematically as ($\Delta t/t_{TRT}$). This proportionality factor is always less than or equal to one, in most cases it dramatically reduces the extra process time to be added for correcting the deviation, and always results in a corrected process that delivers no less than the final target lethality specified for the original process, but with near minimum extended process time. This reduction also brings with it minimal compromise in product quality, as well as minimal disruption of retort scheduling in cook room operations.

The logic behind this "proportional-correction" strategy stems from the following rationale:

- The current industry practice (extending process time to that which would be needed to deliver the same final lethality had the entire process been carried out with an alternative lower constant retort temperature equal to that reached at the lowest point in the deviation) is necessary only when the deviation fails to recover, and retort temperature remains at the lowest point for the duration of the process.
- This practice is unnecessary when the deviation recovers, and processing resumes at the original scheduled retort temperature over the "corrected" extended time, resulting in final lethality far in excess of that required and associated poor quality.

- If the extended process time is chosen to be in proportion to the duration of the deviation as a fraction of original scheduled process time, we are making the assumption that the amount of lethality lost during the deviation duration time is the amount that would have accumulated at retort temperature.
- In reality, this amount of lost lethality is much less, since the actual retort temperature had fallen during the deviation to some lower level where lethality would accumulate at a slower rate.
- Therefore, the "proportional correction" should always deliver total final lethality greater than that originally specified for the scheduled process, while dramatically reducing the degree of unnecessary overprocessing that would otherwise occur with current industry practice.
- With the implementation of this novel and efficient on-line strategy it will be unnecessary for processors to operate at higher retort temperatures.

### *4.3.2. Performance demonstration*

This task consisted of demonstrating the performance of these strategies by simulating the occurrence of process deviations happening at different times during the process (early, late, and randomly) to both solid and liquid canned food products, calculating the alternative corrected process times, and predicting the outcomes of each corrected process in terms of final lethality and quality retention. For each deviation, three different alternative corrected process times were calculated:

1. *Exact correction*—giving corrected process time to reach precisely the final target lethality specified for the scheduled process, using computer simulation with heat transfer models.
2. *Proportional correction*—using the strategy described in this paper with look-up tables.
3. *Commercial correction*—using current industry practice with look-up tables (manually or computerized).

The heat transfer models were explicitly chosen to simulate the two extreme heat transfer cases encountered in thermal processing of canned foods. The rationale behind this decision was that canned foods possess heating characteristics between these two extreme situations. Conclusions extracted from these simulations will be extended to all canned foods. The first case was that of pure conduction heating of a solid product under a still-cook retort process. The second case was that of forced convection heating of a liquid product under mechanical agitation. In both cases, the container shape was assumed to be a finite cylinder, which is typical of a metal can or wide-mouthed glass jar. However, suitable models appropriate for true container shapes can be used as required for this purpose. Examples of such models can be found in the literature (Teixeira *et al.*, 1969; Manson *et al.*, 1970, 1974; Datta *et al.*, 1986; Simpson *et al.*, 1989; Simpson *et al.*, 2004). The product and process conditions chosen to carry out the demonstrated simulations for each case are given in Table 4.3.

### *4.3.3. Demonstration of safety assurance by complex optimization search routine*

This on-line correction strategy was validated and tested for safety assurance by executing a strict and exhaustive search routine with the use of the heat transfer models selected in Task 4.3.2 on high-speed computer. The problem to be solved by the search routine was to determine if the minimum final lethality delivered by all the corrected processes that could be found among the various types of deviations and process conditions considered in the problem domain met the criterion that it had to be greater than or equal to the lethality specified for the original scheduled process. This criterion can be expressed mathematically:

$$\operatorname*{Min}_{U}\left[F_{\text{proportional}} - F_{\text{TOL}}\right] \geq 0$$

where $F_{\text{proportional}}$ is $F$-value calculated with the proportional correction, $F_{\text{TOL}}$ is $F$-value specified for normal scheduled process, and $U$ is universe of feasible process conditions in search routine.

Table 4.4 identifies the various types of deviations and process conditions that were explored and evaluated in the search routine

**Table 4.3.** Product and process conditions used for on-line correction strategy simulations

| | Dimension (cm) | | | Properties | | Normal process | |
|---|---|---|---|---|---|---|---|
| Product simulated | Major | Intermedium | Minor | Alpha ($m^2/s$) | $f_h$ (min) | Time (min) | TRT (°C) |
| Pure conduction can, Biot > 40 | 11.3 | – | 7.3 | $1.70 \times 10^{-7}$ | 44.4 | 64.1 | 120 |
| Forced convection can, Biot < 1 | 11.3 | – | 7.3 | – | 4.4 | 15.6 | 120 |

**Table 4.4.** Problem domain for search routine

| Process variable | Description of process variable | Minimum value | Maximum value |
|---|---|---|---|
| TRT | Scheduled retort temperature | 110°C | 135°C |
| $TRT_i$ | Lowest retort temperature reached during deviation *i* | 100°C | TRT – 0.5°C |
| $t_{CUT}$ | Initial come-up time of retort to reach TRT | 5 minutes | 15 minutes |
| $t_{dev\text{-}i}$ | Time during the process at which the deviation *i* begins | $t_{cut}$ | $t_{TRT}$ |
| $t_i$ | Time duration of the deviation *i* | 0.5 minute | $t_{TRT} - t_{dev\text{-}I}$ |
| $T_{ini}$ | Initial product temperature | 20°C | 70°C |

(problem domain). The search routine was designed as an attempt to find a set of conditions under which the required search constraint was not met. The table gives the symbol used to represent each variable and a description of that variable, along with the minimum and maximum values limiting the range over which the search was conducted.

## 4.4. Equivalent lethality curves

Recall the first requirement to carry out the proportional or commercial on-line correction of a process deviation was to have access to charts or look-up tables that identified the process time needed at different retort temperatures to deliver the same final lethality. These are used to find the alternative process time for the corrected process, and can be presented graphically as "equivalent process lethality curves" for each scheduled product/process. Therefore, equivalent process lethality curves were constructed for each of the two simulated products used in this study, and are shown in Figures 4.5 and 4.6 for the case of solid (pure conduction) and liquid (forced convection), respectively.

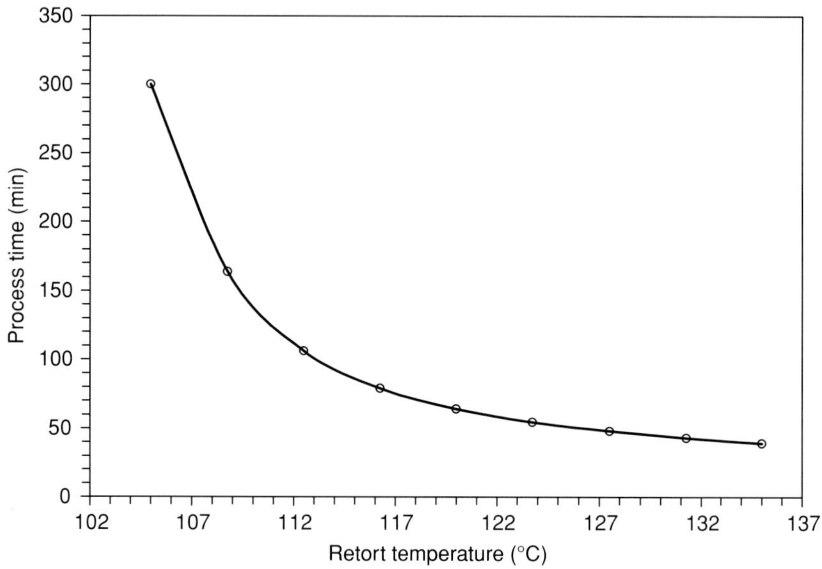


**Figure 4.5.** Equivalent process lethality curve for simulated solid product under pure conduction heating, showing retort temperature/process time combinations that deliver the same final target lethality.

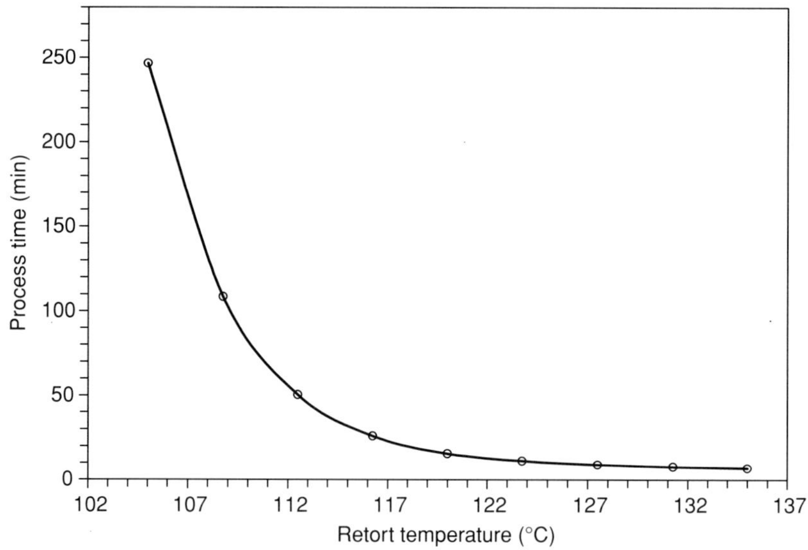


**Figure 4.6.** Equivalent process lethality curve for simulated liquid product under forced convection heating, showing retort temperature/process time combinations that deliver the same final target lethality.

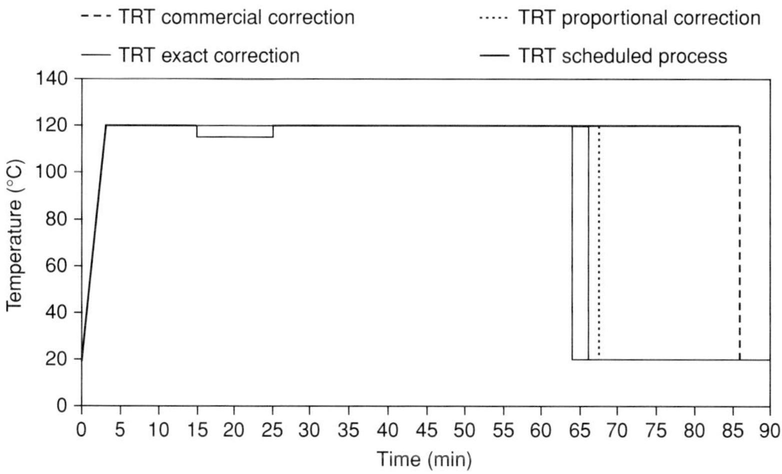


**Figure 4.7.** Pure conduction simulation for on-line correction of an unexpected retort temperature deviation occurring early into the scheduled process for a cylindrical can of solid food under still cook.

### *4.4.1. Performance demonstration*

Figures 4.7–4.10 show results from the four product/process simulations carried out to demonstrate the performance of these strategies. The figures contain retort temperature profiles resulting from on-line

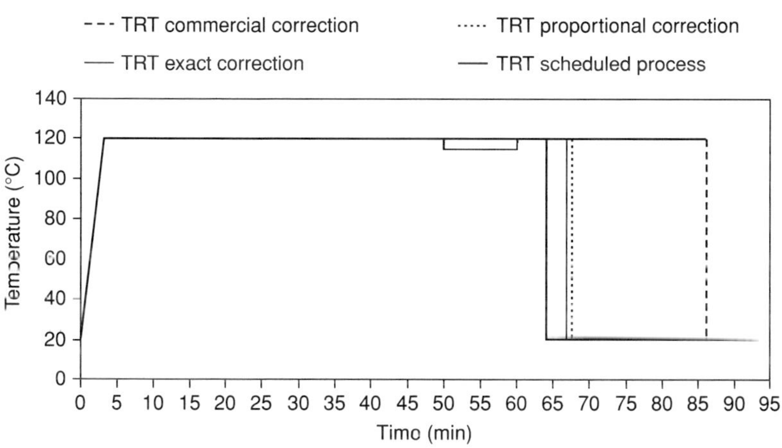


**Figure 4.8.** Pure conduction simulation for on-line correction of an unexpected retort temperature deviation occurring late into the scheduled process for a cylindrical can of solid food under still cook.

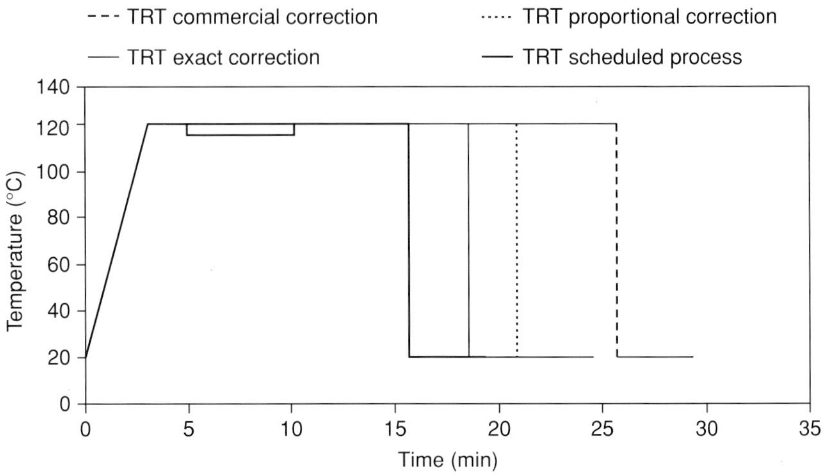


**Figure 4.9.** Forced convection simulation for on-line correction of an unexpected retort temperature deviation occurring early into the scheduled process for a cylindrical can of liquid food under agitated cook.

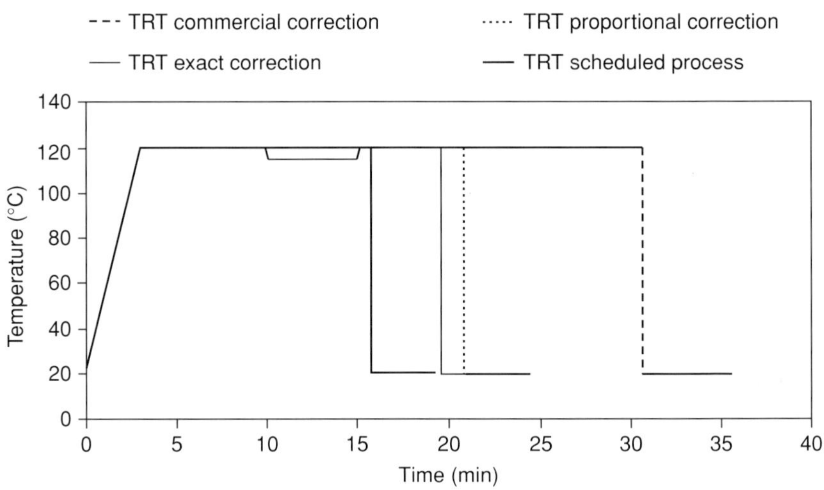


**Figure 4.10.** Forced convection simulation for on-line correction of an unexpected retort temperature deviation occurring late into the scheduled process for a cylindrical can of liquid food under agitated cook.

correction of process deviations happening at different times during the process (early and late) to both solid and liquid canned food products. Each figure shows the "normal" constant retort temperature profile expected for the originally scheduled process, along with the occurrence of a deviation (sudden step drop in retort temperature for short duration) either relatively early or late into the process. In addition, for each deviation (one in each figure), three different alternative corrected process times are shown resulting from different strategies:

1. *Exact correction*—giving corrected process time to reach precisely the final target lethality specified for the scheduled process, using computer simulation with heat transfer models.
2. *Proportional correction*—using the new strategy with equivalent lethality curves; and
3. *Commercial correction*—using current industry practice with the same equivalent lethality curves.

In all cases, the extended process time required by the "commercial-correction" strategy, which is used now in current industry practice, is far in excess of the extended times called for by the other two strategies. Moreover, the new "proportional-correction" strategy results in extending process time only slightly beyond that required for an "exact correction," and will always do so. These results are summarized in Table 4.5, along with results from predicting the outcomes of each corrected process in terms of final process times required, and final lethality and quality retention achieved (using product/process data presented in Table 4.3). It is most interesting to note the dramatic improvement in nutrient (quality) retention between that resulting from the commercial correction (current industry practice) and that resulting from either of the other two strategies.

### *4.4.2. Demonstration of safety assurance by complex search routine*

This on-line correction strategy was validated and tested for safety assurance by executing an exhaustive search routine with the use of the heat transfer models. Recall the problem to be solved by the search

**Table 4.5.** Outcomes of each corrected process deviation described in Figures 4.7–4.10 in terms of final process time, lethality, and quality retention for the three different alternative correction methods

| | Early deviation | | | Late deviation | | |
|---|---|---|---|---|---|---|
| | Time (min) | $F_o$ (min) | Nutrient retention (%) | Time (min) | $F_o$ (min) | Nutrient retention (%) |
| Pure conduction | | | | | | |
| Scheduled process | 64.1 | 6.0 | 72.7 | 64.1 | 6.0 | 72.7 |
| Exact correction | 66.3 | 6.0 | 72.9 | 66.8 | 6.0 | 72.7 |
| Proportional correction | 67.5 | 6.5 | 72.2 | 67.5 | 6.2 | 72.3 |
| Comercial correction | 86.2 | 16.3 | 62.3 | 86.2 | 14.4 | 62.8 |
| Foreced convection | | | | | | |
| Scheduled process | 15.6 | 6.0 | 92.4 | 15.6 | 6.0 | 92.4 |
| Exact correction | 18.4 | 6.1 | 91.5 | 19.6 | 6.0 | 90.9 |
| Proportional correction | 20.8 | 8.0 | 89.7 | 20.8 | 7.0 | 89.9 |
| Comercial correction | 25.6 | 11.8 | 86.2 | 30.6 | 14.7 | 82.8 |

routine was to determine if the minimum final lethality delivered by all the corrected processes that could be found among all the various types of deviations and process conditions considered in the problem domain met the criterion that it had to be greater than or equal to the lethality specified for the original scheduled process. Table 4.4 (presented earlier) identifies the problem domain by specifying the various types of deviations and process conditions that were explored and evaluated in the search routine. The search routine was designed to find a set of conditions under which the required search constraint was not met. No such conditions could be found.

## 4.5. Industrial automation of batch retorts

Many of the most recent advances made in the design of industrial batch retorts has come about in response to the increasing popularity of flexible retort pouches and retortable semi-rigid microwavable plastic dinner trays and lunch bowls. These flexible and semi-rigid containers lack the strength of traditional metal cans and glass jars to withstand the large pressure differences experienced across the container during normal retort operations. To safely process these types of flexible packages, careful control of overriding air pressure is needed during retort processing, and pure saturated steam, alone, cannot be used as the heat exchange medium. Instead, new retorts designed to be used with pressure-controlled steam–air mixtures, water spray, or water cascade have been recently developed for this purpose (Blattner, 2004). Examples of some of these new retort designs are given in Figure 4.11. A close-up view of some of the specially designed racking configurations used to hold flexible retortable packages in place during retorting is shown in Figure 4.12.

Perhaps the most significant advances made in the food canning industry to date have been in the area of automated materials handling systems for loading and unloading batch retorts. Traditionally, the loading and unloading of batch retorts has been the most labor-intensive component in food canning factories. Unprocessed sealed containers would be manually stacked into baskets, crates, or carts. Then, the baskets or crates would be loaded into empty vertical retorts with the aid of chain hoist, or wheeled carts

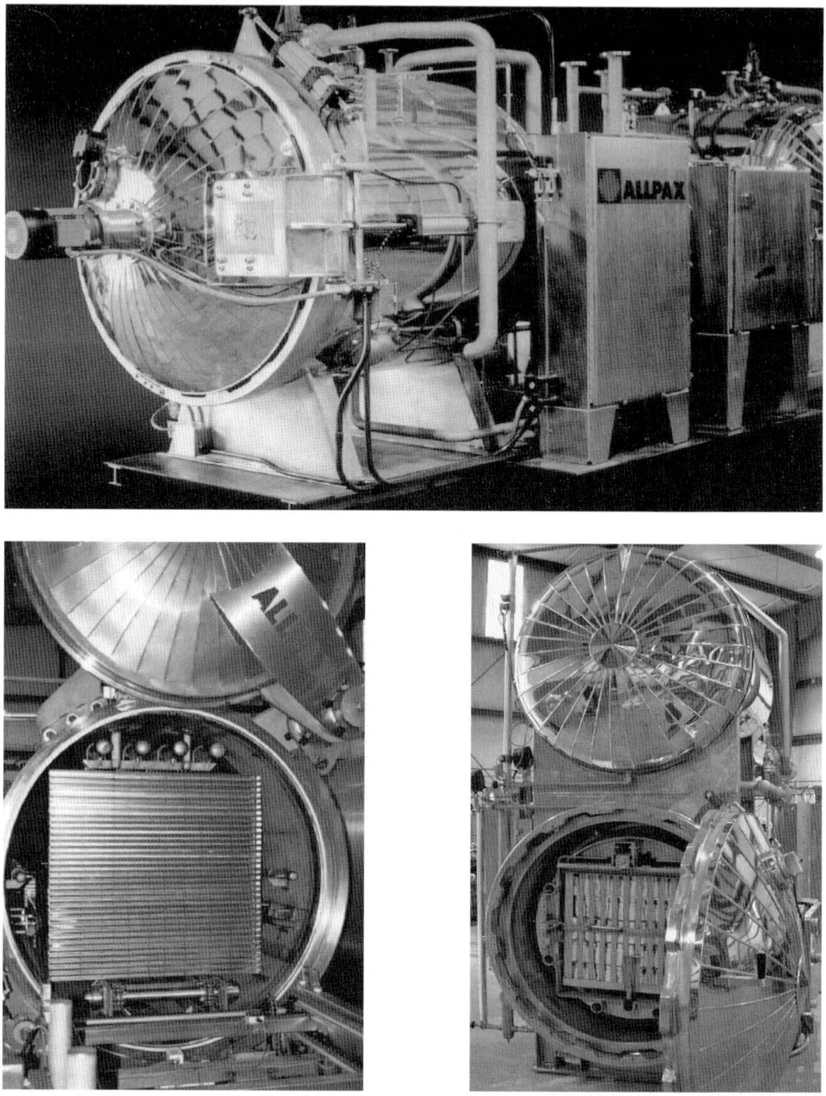


**Figure 4.11.** New retort systems (rotating and still cook) with specially designed racking configurations for processing flexible and semirigid packages. (Courtesy of ALLPAX, Covington, LA.)

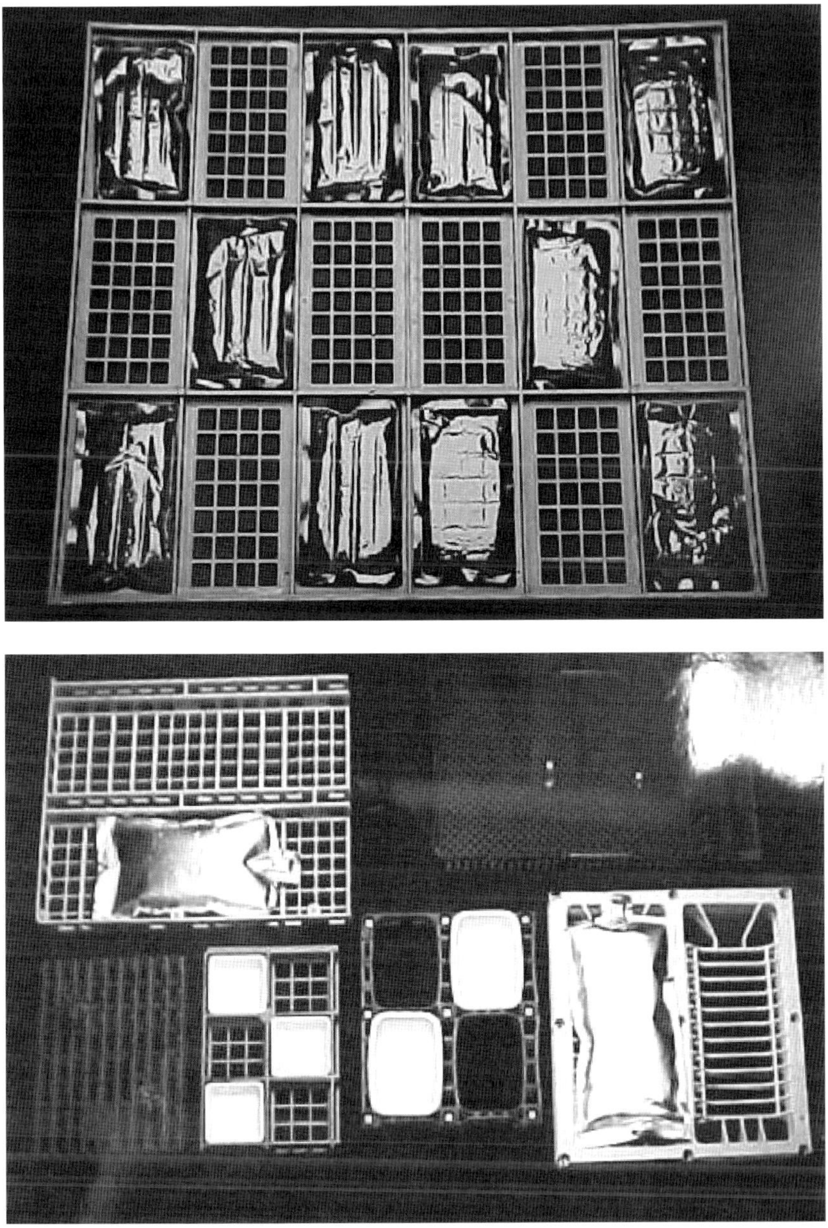

**Figure 4.12.** Rack designs for flexible and semi-rigid retortable packaging systems. (Courtesy of ALLPAX, Covington, LA.)

would be loaded into horizontal retorts with the aid of track rails for this purpose. In recent years, leading manufacturers of retort systems have been offering a host of new automated materials handling systems to automate the retort loading and unloading operation.

Most of the new automated systems available to date are based on the use of either automated guided vehicles (Heyliger, 2004) or orthogonal direction shuttle systems (Blattner, 2004; Heyliger, 2004). Both types of systems are designed for use with horizontal retorts. The automated guided vehicles (AGVs) works like robots. They carry the loaded crates of unprocessed product from the loading station to any designated retort on the cook room floor that is ready to be loaded. They also carry the loaded crates of finished processed product from the unloaded retort to the unloading station for discharge as outgoing product exiting the cook room to the case packing operation. These robotic AGVs are designed to integrate with the loading station in such a way that sealed product containers arriving on a conveyor automatically stack into the crate carried by the AGV, which later inserts the entire crate into the designated retort. Unloading at the unloading station for finished product discharge is likewise accomplished in a similar automated way, but in reverse. The AGVs are guided by an underground wire tracking system buried beneath the cook room floor. This leaves the cook room floor space open and free of any rail tracks or guide rails that would otherwise impede the safe movement of factory workers in their normal work flow operations. A panoramic view of a large cook room operation using an automated batch retort system with AGVs is shown in Figure 4.13 (Heyliger, 2004), and a close-up view of an AGV in the process of loading or unloading a horizontal retort is shown in Figure 4.14.

An alternative to the AGV system is the shuttle system offered by several retort manufactures. Unlike the AGV system, the shuttle system relies upon a set of tracks or rails that are fixed in place on the cook room floor. These rails span the length of the cook room along the row of horizontal retorts, allowing a shuttle carrying loaded crates to slide along these rails until it has aligned itself in front of the designated retort waiting to be loaded. In a similar fashion, when a retort is ready for unloading, an empty shuttle slides along these rails until it has aligned itself with that retort to receive the loaded crates of processed product. Then the shuttle slides along the rails to

**Figure 4.13.** Automated batch retort system with use of automated guided vehicles in large cook room operation. (Courtesy of JBT FoodTech, formerly FMC FoodTech, Madera, CA.)

**Figure 4.14.** Automated guided vehicle for batch retort loading/unloading. (Courtesy of JBT FoodTech, formerly FMC FoodTech, Madera, CA.)

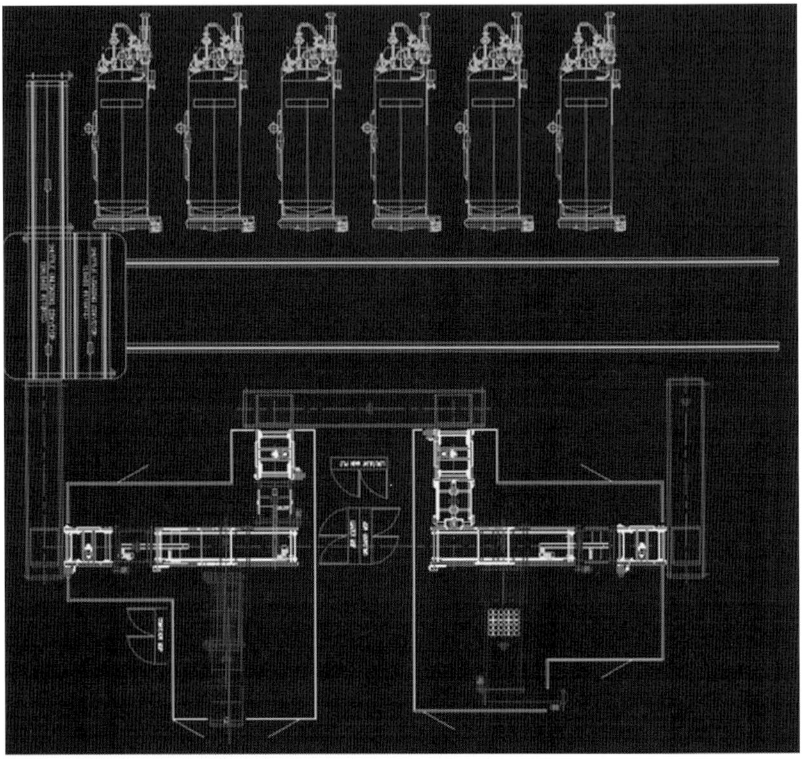

**Figure 4.15.** Automated shuttle-based batch retort control system. (Courtesy of ALLPAX, Covington, LA.)

far end of the cook room where unloading of processed product takes place for discharge out of the cook room. Normally, the unprocessed product loading station and the processed product unloading stations are located at opposite ends of the cook room (Figure 4.15). Figures 4.16 and 4.17 illustrate the shuttle systems offered by ALLPAX and FMC, respectively.

## 4.6. Concluding remarks

This chapter described the development of broader applications of computer-based batch retort control systems that make use of mathematical models capable of accomplishing on-line correction of

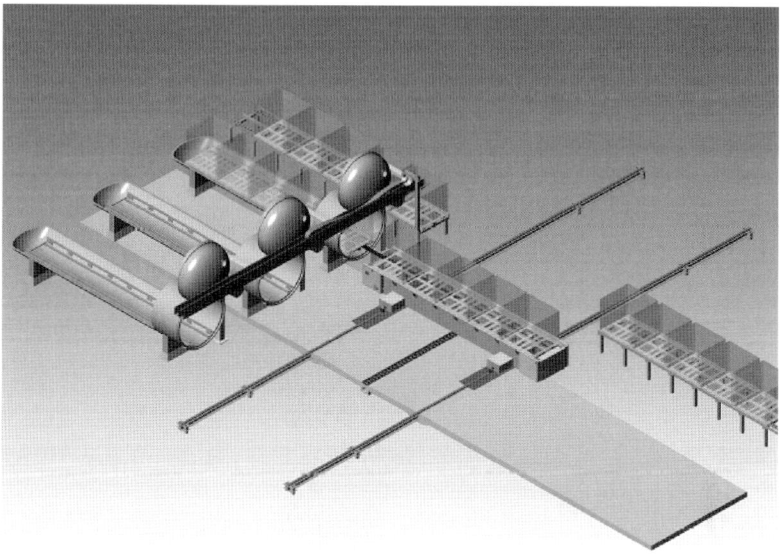

**Figure 4.16.** Automated shuttle batch retort system. (Courtesy of ALLPAX, Covington, LA.)

**Figure 4.17.** FMC shuttle system for automated batch retort loading/unloading. (Courtesy of JBT FoodTech, formerly FMC FoodTech, Madera, CA.)

unexpected process deviations in thermal processing of low-acid canned foods. It specifically addressed the problem posed by current systems that automatically extend process time at the recovered retort temperature for the extent needed to reach the target lethality ($F_0$) specified for the product.

### *4.6.1. On-line control strategies applicable to low-acid foods that can be mathematically modeled*

The work presented in this paper addressed these problems by describing novel control strategies that also treat the retort temperature as the control variable, rather than process time alone, in accomplishing on-line correction of a process deviation. In the case of cook room scheduling situations, on-line correction of a deviation is accomplished by choosing an optimum higher constant retort temperature for the remainder of the process that will deliver the specified target lethality within the original process time remaining. When the deviation occurs late in the process requiring correction at the upper limit of retort temperature, this would be a minimum extended process time. In a further and even broader application, the paper described on-line correction of process deviations occurring during a preprogrammed variable or dynamic retort temperature process, such as a ramp-up and ramp-down profiles, that might be chosen to maximize nutrient retention in a conduction-heated food. In this situation, an optimum combination of retort temperature and process time is chosen for the remainder of the process that will deliver the maximum possible nutrient retention within the specified target lethality. Examples were given for the case of solid product undergoing a conduction-heating process in different shaped containers (cylindrical can and retort pouch), as well as liquid product undergoing a forced convection-heating process in cylindrical cans under mechanical agitation.

### *4.6.2. On-line control strategies applicable to any low-acid foods*

The strategy is intended for easy implementation in any cannery around the world with no need for on-site access to computers, computer-based control systems, and/or computer software of any kind. This strategy takes into account the duration of the deviation in addition to the magnitude of the temperature drop. It calculates a “proportional” extended process time at the recovered retort temperature that will deliver the final specified target lethality with very little overprocessing in comparison to current industry practice. Results from an exhaustive search routine using the complex method support the logic and rationale behind the strategy by showing that the proposed strategy will always result in a corrected process that delivers no less than the final target lethality specified for the originally scheduled process.

## List of Symbols

| | |
|---|---|
| $A$ | Area |
| $a$ and $b$ | Constant of linear equation describing retort temperature profile ($\mathrm{TRT}(t) = a + bt$) |
| $b'$ | New slope of the linear equation describing retort temperature profile ($\mathrm{TRT}'(t) = a' + b't$) |
| $a'$ | New constant of the linear equation describing retort temperature profile ($\mathrm{TRT}'(t) = a' + b't$) |
| $C_p$ | Heat capacity of food |
| CUT | Come-up time |
| $\underline{E}$ | Energy per mass unit |
| $F_o$ | Sterilizing value at 121.1°C |
| $F_{\mathrm{proportional}}$ | $F$-value calculated with the proportional correction. |
| $F_{\mathrm{TOL}}$ | $F$-value specified for normal scheduled process |
| $f$ | Rate factor (related to slope of semi-log heat penetration curve) |
| $f_h$ and $f_c$ | Heating and cooling rate factors (related to slope of semi-log heat penetration curve) |
| $j$ | Dimensionless lag factor ($j = \frac{TRT - T_A}{TRT - IT}$) |
| $j_h$ and $j_c$ | Heating and cooling lag factors |
| $k$ | Thermal conductivity of food |
| $l$ | Height of canned content |
| $M$ | Product mass |
| $n$ | Number of deviations occurring during the process. |
| TRT | Retort temperature |

| | |
|---|---|
| TRT′ | New retort temperature |
| $TRT_i$ | Lowest temperature during the deviation *i* |
| $t$ | Time |
| $t_D$ | Corrected process time |
| $t_{TRT}$ | Preestablished process time at retort temperature TRT |
| $U$ | Global heat transfer coefficient |
| $\Delta t_i$ | Duration of deviation *i* |

## Greek letters

| | |
|---|---|
| $\alpha$ | Thermal diffusivity of food ($\alpha = k/\rho C_p$) |
| $\rho$ | Density of food |
| $\nabla$ | Differential or nabla operator ($\nabla = \partial/\partial x + \partial/\partial y + \partial/\partial z$) |
| $\nabla^2$ | Laplace operator ($\nabla^2 = \partial^2/\partial x^2 + \partial^2/\partial y^2 + \partial^2/\partial z^2$) |

## Acknowledgment

Ricardo Simpson is grateful for the financial support provided by CONICYT through the FONDECYT project 1050810.

## References

Alonso, A., Banga, J., and Perez-Martin, R. 1993. A new strategy for the control of pressure during the cooling stage of the sterilization process in steam retorts. Part I. A preliminary study. Food and Bioproducts Processing, *Trans IchemE*, 71(c): 197–205.

Ball, C.O. 1928. Mathematical solution of problems on thermal processing of canned food. University of California. Published in Public Health 1, N 2, 15–245.

Bigelow, W.D., Bohart, G.S., Richardson, A.C. and Ball, C.O. 1920. Heat penetration in processing canned foods. Bull. No. 16-L Res. Lab. Natl. Canners Assn., Washington, DC.

Blattner, M. F. 2004. Advances in automated retort control, and today's new packaging. Presentation at IFT Symposium, 2004 IFT Meeting, Las Vegas, NV.

Datta, A.K., Texeira, A.A., and Manson, J.E. 1986. Computer-based retort control logic for on-line correction of process deviations. *Journal of Food Science*, 51(2): 480–483, 507.

Heyliger, T. L. (2004). Advances in retort control for batch and continuous systems. Presentation at IFT Symposium, 2004 IFT Meeting, Las Vegas, NV.

Holdsworth, S.D. 1997. *Thermal Processing of Packaged Foods*. Blackie Academic & Professional, London.

Larkin. 2002. Personal Communication. Branch Chief, National Center for Food Safety and Technology, Food and Drug Administration (FDA/NCFST), Chicago, IL.

Manson, J.E., Zahradnik, J.W., and Stumbo, C.R. 1970. Evaluation of lethality and nutrient retentions of conduction-heating food in rectangular containers. *Food Technology*, 24(11): 1297–1301.

Manson, J.E., Zahradnik, J.W., and Stumbo, C.R. 1974. Evaluation of thermal processes for conduction heating foods in pear-shaped containers. *Journal of Food Science*, 39: 276–281.

Noronha, J., Hendrickx, M., Van Loey, A., and Tobbak, P. 1995. New semi-empirical approach to handle time-variable boundary conditions during sterilization of non-conductive heating foods. *Journal of Food Engineering*, 24: 249–268.

Pham, Q.T. 1987. Calculation of thermal process lethality for conduction-heated canned foods. *Journal of Food Science*, 52(4): 967–974.

Saguy, I. and Karel, M. 1979. Optimal retort temperature profile for optimizing thiamine retention in conduction-type heating canned foods. *Journal of Food Science*, 44, 1485–1490.

Simpson, R., Aris, I., and Torres, J.A. 1989. Sterilization of conduction-heated foods in oval-shaped containers. *Journal of Food Science*, 54(5): 1327–1331, 1363.

Simpson, R. 2004. Control logic for on-line correction of batch sterilization processes applicable to any kind of canned food. Symposium of Thermal processing in the 21st century: Engineering modelling and automation, at IFT Meeting, 2004, Las Vegas, NV.

Simpson, R., Mitchell, M. and Almonacid, S. 2004. Mathematical model development, experimental validation and process optimization: retortable pouches packed with seafood in cone frustum shape. *Journal of Food Engineering*, 63(2): 153–162.

Stumbo, C.R. 1973. *Thermobacteriology in Food Processing*, 2nd ed. Academic Press, New York.

Teixeira, A., Dixon, J., Zahradnik, J., and Zinsmeiter, G. 1969. Computer optimization of nutrient retention in the thermal processing of conduction-heated foods. *Food Technology*, 23(6): 845–850.

Teixeira, A.A., Zinsmeister, G.E., and Zahradnik, J.W. 1975. Computer simulation of variable retort control and container geometry as a possible means of improving thiamine retention in thermally-processed foods. *Journal of Food Science*, 40(3): 656–659.

Teixeira, A.A. 1992. Thermal process calculations. In *Handbook of Food Engineering*, D.R. Heldman and D.B. Lund (eds.). Marcel Dekker, Inc., New York, pp. 563–619.

Teixeira, A.A. and Manson. J.E. 1982. Computer control of batch retort operations with on-line correction of process deviations. *Food Technology*, 36(4): 85–90.

Teixeira, A.A. and Tucker, G.S. 1997. On-line retorts control in thermal sterilization of canned foods. *Food Control*, 8(1): 13–20.

Teixeira, A.A., Balaban, M.O., Germer, S.P.M., Sadahira, M.S., Teixeira-Neto, R.O., and Vitali, A.A. 1999. Heat transfer model performance in simulating process deviations. *Journal of Food Science*, 64(3): 488–493.

Von Oetinger, K. 1997. Lógica para el control en línea del proceso de esterilización comercial. Tesis Escuela de Alimentos. Pontificia Universidad Católica de Valparaíso, Valparaíso, Chile.

# Chapter 5

# COMPUTER SOFTWARE FOR ON-LINE CORRECTION OF PROCESS DEVIATIONS IN BATCH RETORTS

*Arthur A. Teixeira and Murat O. Balaban*

## 5.1. Introduction

Thermal processing of canned foods has been one of the most widely used methods of food preservation during the twentieth century and has contributed significantly to the nutritional well-being of much of the world's population. Thermal processing consists of heating food containers in pressurized retorts at specified temperatures for prescribed lengths of time. These process times are calculated on the basis of achieving sufficient bacterial inactivation in each container to comply with public health standards and to ensure that the probability of spoilage will be less than some minimum. Associated with each thermal process is always some degradation of heat-sensitive vitamins and other quality factors that is undesirable. Because of these quality and safety factors, great care is taken in the calculation of these process times and in the control of time and temperature during processing to avoid either under- or overprocessing. The heat transfer considerations that govern the temperature profiles achieved within the container of food are critical factors in the determination of time and temperature requirements for sterilization. This chapter will

*Thermal Processing of Foods: Control and Automation.* Edited by K.P. Sandeep

focus on the development and application of deterministic heat transfer models capable of accurately predicting internal product temperature in response to retort operating conditions, and coupling these with deterministic models that mathematically describe the thermal inactivation kinetics of bacterial spores and food quality factors for thermal process simulation.

## 5.2. Thermal death time relationships

An understanding of two distinct bodies of knowledge is required to appreciate the basic principles involved in thermal process calculation. The first of these is an understanding of the thermal inactivation kinetics (heat resistance) of food-spoilage-causing organisms. The second body of knowledge is an understanding of heat transfer considerations that govern the temperature profiles achieved within the food container during the process, commonly referred to in the canning industry as heat penetration.

Figure 5.1 conceptually illustrates the interdependence between the thermal inactivation kinetics of bacterial spores and the heat transfer considerations in the food product. Thermal inactivation of bacteria generally follows first-order kinetics and can be described by logarithmic reduction in the concentration of bacterial spores with time for any given lethal temperature, as shown in the upper family of curves in Figure 5.1. These are known as *survivor curves*. The decimal reduction time, $D$, is expressed as the time required to achieve one log cycle of reduction in concentration, $C$. As suggested by the family of curves shown, $D$ is temperature dependent and varies logarithmically with temperature, as shown in the second graph. This is known as a thermal death time curve and is essentially a straight line over the range of temperatures employed in food sterilization. The slope of the curve that describes this relationship is expressed as the temperature difference, $Z$, required for the curve to transverse one log cycle. The temperature in the food product, in turn, is a function of the retort temperature ($T_R$), initial product temperature ($T_I$), location within the container ($x$), thermal diffusivity of the product ($\alpha$), and time ($t$) in the case of a conduction-heating food.

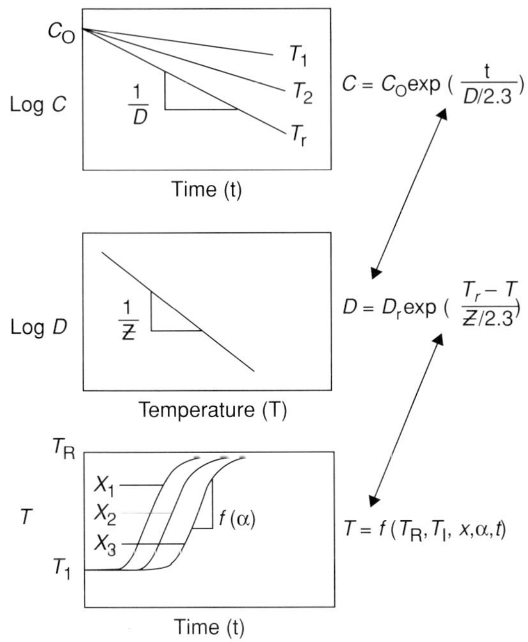


**Figure 5.1.** Time and temperature dependence of thermal inactivation kinetics of bacterial spores in thermal processing of canned foods.

Thus, the concentration of viable bacterial spores during thermal processing decreases with time in accordance with the inactivation kinetics, which are a function of temperature. The temperature, in turn, is a function of the heat transfer considerations involving time, spatial location, combined thermal and physical properties (thermal diffusivity), and initial and boundary conditions (initial product temperature and retort temperature, respectively).

## 5.3. Process lethality and sterilizing value

### *5.3.1. Time at temperature for isothermal process*

Once the thermal death time (TDT) curve (Figure 5.2) has been established for a given microorganism, it can be used to calculate the time–temperature requirements for any idealized thermal process

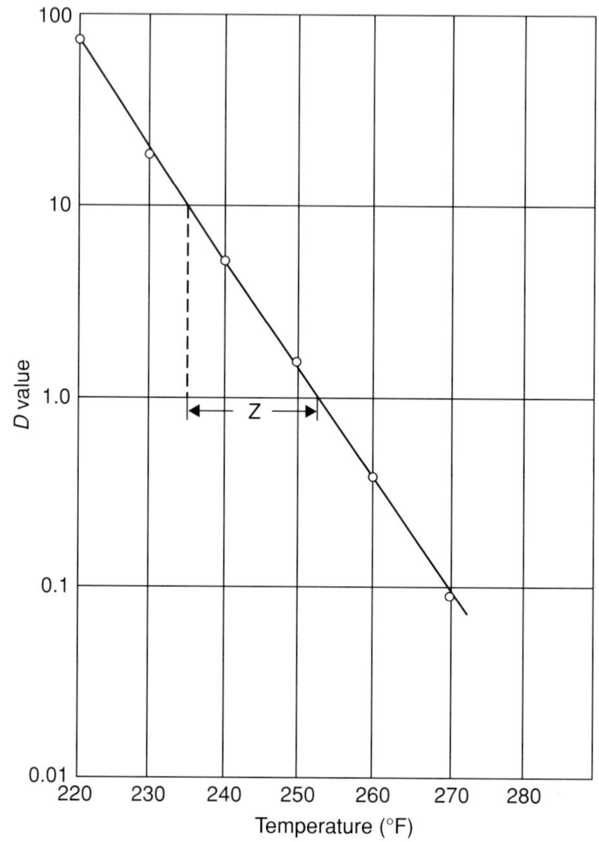


**Figure 5.2.** Thermal death time (TDT) curve showing temperature dependency of *D*-value given by temperature change (*Z*) required for tenfold change in *D*-value.

(isothermal process) in which product is heated instantly and uniformly to the treatment temperature, held there for a specified time, and likewise cooled instantly and uniformly. For example, assume a process is required that will achieve a six-log-cycle reduction in the population of bacterial spores whose kinetics are described by the TDT curve in Figure 5.2 at a specified process temperature ($T$). The $D$-value at that temperature is taken from the curve and simply multiplied by the number of log cycles of spore reduction required to determine the process time needed.

Since the TDT curve is a straight line on a semi-log plot, all that is needed to specify such a curve is its slope and a single reference point on the curve. The slope of the curve is specified by the *Z*-value, and the reference point is the *D*-value at a reference temperature. For sterilization of low-acid foods (pH above 4.5), in which thermophilic spores of relatively high heat resistance are of concern, this reference temperature is usually taken to be 121.1°C (250°F). For high-acid foods or *pasteurization* processes in which microorganisms of much lower heat resistance are of concern, lower reference temperatures are used, such as 100°C or 66°C. In specifying a reference *D*-value for a microorganism, the reference temperature is shown as a subscript, such as $D_{121.1}$. Ranges of *D*-values for different classification of bacteria are given in Table 5.1, and $D_{121.1°C\ (250°F)}$-values for specific organisms in selected food products are given in Table 5.2.

**Table 5.1.** *D*-values for different classifications of foodborne bacteria

| Bacterial groups | *D*-value |
|---|---|
| Low-acid and semiacid foods (pH above 4.5) | $D_{250}$ |
| Thermophiles | |
| Flat-sour group (*Bacillus stearothermophilus*) | 4.0–5.0 |
| Gaseous-spoilage group (*Clostridium thermosaccharolyticum*) | 3.0–4.0 |
| Sulfide stinkers (*Clostridium nigrificans*) | 2.0–3.0 |
| Mesophiles | |
| Putrefactive anaerobes | |
| *C. botulinum* (types A and B) | 0.10–1.20 |
| *C. sporogenes* group (including P.A. 3679) | 0.10–1.5 |
| Acid foods (pH 4.0–4.5) | |
| Thermophiles | |
| *Bacillus coagulans* (facultatively mesophilic) | 0.01–0.07 |
| Mesophiles | $D_{212}$ |
| *Bacillus polymyxa* and *Bacillus macerans* | 0.10–0.50 |
| Butyric anaerobes (*Clostridium pasteurianum*) | 0.10–0.50 |
| High-acid foods (mesophilic non-spore-bearing bacteria) | $D_{150}$ |
| *Lactobacillus* spp., *Leuconostoc* spp., yeast and molds | 0.50–1.00 |

*Source*: Stumbo (1965).

**Table 5.2.** Comparison of $D_{121.1^{\circ}C\ (250^{\circ}F)}$-values for specific microorganisms in selected food substrates

| Organism | Substrate | TDT method | $D_{250}$ (minutes) |
|---|---|---|---|
| P.A. 3679 | Cream-style corn | Can | 2.47 |
| P.A. 3679 | Whole-kernel corn | Can | 1.52 |
| P.A. 3679 | Whole-kernel corn (replicate) | Can | 1.82 |
| P.A. 3679 | Phosphate buffer | Tube | 1.31 |
| F.S. 5010 | Cream-style corn | Can | 1.14 |
| F.S. 5010 | Whole-kernel corn | Can | 1.35 |
| F.S. 1518 | Phosphate buffer | Tube | 3.01 |
| F.S. 617 | Whole milk | Can | 0.84 |
| F.S. 617 | Evaporated milk | Tube | 1.05 |

*Source*: Stumbo, 1965.
P.A., putrefactive anaerobe; F.S., facultative spore.

### 5.3.2. Process lethality

The example process calculations carried out in the preceding subsection show that, for a given $Z$-value, the specification of any one point on the straight line drawn parallel to the TDT curve but intersecting the process time at process temperature is sufficient to specify the sterilizing value of any process combination of time and temperature on that line. The reference point that has been adopted for this purpose is the time in minutes at the reference temperature of 121.1°C, or the point in time where the equivalent process curve crosses the vertical axis drawn at 121.1°C, and is known as the $F$-value for the process. This is often referred to as the lethality of a process, and since it is expressed in minutes at 121.1°C, the unit of lethality is 1 minute at 121.1°C. Thus, if a process is assigned an $F$-value of 6, it means that the integrated lethality achieved by whatever time–temperature history is employed by the process must be equivalent to the lethality achieved from 6 minutes exposure to 121.1°C, assuming an idealized process of instantaneous heating to 121.1°C followed by instantaneous cooling after the 6-minute hold.

All that is required to specify the $F$-value is to determine how many minutes at 121.1°C will be required to achieve the specified level of log-cycle reduction. The $D_{121.1^{\circ}C}$-value is used for this purpose, because it represents the number of minutes at 121.1°C to accomplish one-log-cycle reduction. Thus, the $F$-value is equal to $D_{121}$ multiplied by the sterilizing value (number of log cycles required in population reduction).

$$F = D_{121.1}(\log a - \log b) \tag{5.1}$$

where $a$ is the initial number of viable spores and $b$ is the final number of viable spores (or survivors).

In the example given earlier, assume the value, $D_{121.1} = 1.5$ minutes, was taken from the TDT curve in Figure 5.2, and multiplied by the required sterilizing value (six log cycles). Thus $F = 1.5\ (6) = 9$ minutes, and the lethality for this process has been specified as $F =$ 9 minutes. This is normally the way in which a thermal process is specified for subsequent calculation of a process time at some other temperature. In this way proprietary information regarding specific microorganisms of concern and/or numbers of log cycles reduction can be kept confidential, and replaced by the $F$-value (lethality) as a process specification.

Note also that this $F$-value serves as the reference point to specify the equivalent process design curve discussed earlier. By plotting a point at 9 minutes on the vertical line passing through 121.1°C on a TDT graph, and drawing a line parallel to the TDT curve through this point, the line will pass through all combinations of process time and temperature that deliver the same level of *lethality*. The equation for this straight line can be used to calculate the process time ($t$) at some other constant temperature ($T$) when $F$ is specified.

$$F = 10^{[(T-121.1)/Z]t} \tag{5.2}$$

The following equation becomes important in the general case when the product temperature varies with time during a process, and the $F$-value delivered by the process must be integrated mathematically,

$$F = \int_0^t 10^{[(T-121.1)/Z]dt} \tag{5.3}$$

At this point equations (5.1) and (5.3) have been presented as two clearly different mathematical expressions for the process lethality, $F$. It is most important that the distinction between these two expressions be clearly understood. Equation (5.1) is used to determine the $F$-value that should be *specified* for a process, and is determined from the log-cycle reduction in spore population required of the process (sterilizing value) by considering factors related to safety and wholesomeness of the processed food, as discussed in the following section. Equation (5.3) is used to determine the $F$ *delivered* by a process as a result of the time–temperature history experienced by the product during the process. Another observation is that equation (5.1) makes use of the $D_{121}$-value in converting log cycles of reduction into minutes at 121.1°C, while equation (5.3) makes use of the $Z$-value in converting temperature–time history into minutes at 121.1°C. Because a $Z$-value of 10°C (18°F) is so commonly observed or assumed for most thermal processing calculations, $F$-values calculated with a $Z$ of 10°C and reference temperature of 121.1°C are designated $F_o$.

### 5.3.3. Specification of process lethality

Establishing the sterilizing value to be specified for a low-acid canned food is undoubtedly one of the most critical responsibilities taken on by a food scientist or engineer acting on behalf of a food company in the role of a competent thermal processing authority. In this section we outline briefly the steps normally taken for this purpose (Stumbo, 1965). There are two types of bacterial populations of concern in canned food sterilization. First is the population of organisms of public health significance. In low-acid foods with pH above 4.5, the chief organism of concern is *Clostridium botulinum*. A safe level of survival probability that has been accepted for this organism is $10^{-12}$, or one survivor in $10^{12}$ cans processed. This is known as the 12-D concept for botulinum cook. Since the highest $D_{121}$-value known for this organism in foods is 0.21 minute, the minimum lethality value for a botulinum cook assuming an initial spore load of one organism per container is

$$F = 0.21 \times 12 = 2.52 \text{ min}$$

Essentially, all low-acid foods are processed far beyond the minimum botulinum cook in order to avoid economic losses from spoilage-causing bacteria of much greater heat resistance (the second type). For these organisms, acceptable levels of spoilage probability are usually dictated by marketing and/or economic considerations. Most food companies accept a spoilage probability of $10^{-5}$ from mesophilic spore-formers (organisms that can grow and spoil food at room temperature, but are nonpathogenic). The organism most frequently used to characterize this classification of food spoilage is a strain of *Clostridium sporogenes*, a putrefactive anaerobe (PA) known as PA 3679, with a maximum $D_{121}$-value of 1 minute. Thus, a minimum lethality value for a mesophilic spoilage cook assuming an initial spore of load of one spore per container is

$$F = 1.00 \times 5 = 5.00\,\text{min}$$

Where thermophilic spoilage is a problem, more severe processes may be necessary because of the high heat resistance of thermophilic spores. Fortunately, most thermophiles do not grow readily at room temperature and require incubation at unusually high storage temperatures (45–55°C) to cause food spoilage. Generally, foods with no more than 1% spoilage (spoilage probability of $10^{-2}$) upon incubation after processing, will meet the accepted $10^{-5}$ spoilage probability in normal commerce. Therefore, when thermophilic spoilage is a concern, the target value for the final number of survivors is usually taken as $10^{-2}$, and the initial spore load needs to be determined through microbiological analysis because contamination from these organisms varies greatly. For a situation with an initial thermophilic spore load of 100 spores per can, and an average $D_{121.1}$-value of 4.00, the process *lethality* required would be

$$F = 4.00(\log 100 - \log 0.01) = 4.00(4) = 16\,\text{min}$$

The procedural steps above are only preliminary guidelines for average conditions, and often need to be adjusted up or down in view of the types of contaminating bacteria that may be present, the initial level of contamination or bioburden of the most resistant types, the

**Table 5.3.** Lethality values ($F_o$) for commercial sterilization of selected canned food products

| Product | Can sizes | $F_o$ (minutes) |
|---|---|---|
| Asparagus | | 2 |
| Green beans, brine packed | No. 2 | 3.5 |
| | No. 10 | 3.5 |
| Chicken, boned | All | 6–8 |
| Corn, whole kernel, brine packed | All | 9 |
| | No. 10 | 15 |
| Cream style corn | No. 2 | 5–6 |
| | No. 10 | 2.3 |
| Dog food | No. 2 | 12 |
| | No. 10 | 6 |
| Mackerel in brine | 301 × 411 | 2.9–3.6 |
| Meat loaf | No. 2 | 6 |
| Peas, brine packed | No. 2 | 7 |
| | No. 10 | 11 |
| Sausage, Vienna, in brine | Various | 5 |
| Chili con carne | Various | 6 |

*Source*: Lopez, 1987; courtesy of American Can Company, Inc.

spoilage risk accepted, and the nature of the food product from the standpoint of its ability to support the growth of the different types of contaminating bacteria that are found. Table 5.3 contains a listing of process lethalities ($F_o$) specified for the commercial processing of selected canned foods (NFPA, 1980).

## 5.4. Heat transfer considerations

### *5.4.1. Unsteady (nonisothermal) heat transfer*

In the previous sections on thermal inactivation kinetics of bacterial spores, frequent reference was made to an idealized process in which the food product was assumed to be heated instantaneously to a lethal temperature, then cooled instantaneously after the required process time temperature. These idealized processes are important to gain an understanding of how the kinetic data can be used directly to determine the process time at any given lethal temperature. There are in fact, commercial sterilization processes for which this method of process-time determination is applicable. These are high temperature short time (HTST) pasteurization or ultra-high temperature (UHT) sterilization processes for liquid foods that make use of flow-through heat exchangers and/or steam injection heaters and flash cooling chambers for instantaneous heating and cooling. The process time is accomplished through the residence time in the holding tube between the heater and cooler as the product flows continuously through the system. This method of product sterilization is most often used with aseptic filling systems, discussed in other chapters.

In traditional thermal processing of most canned foods, the situation is quite different from the idealized processes described above. Cans are filled with relatively cool unsterile product, sealed after headspace evacuation, and placed in steam retorts, which apply heat to the outside can wall. The product temperature can then only respond in accordance with the physical laws of heat transfer, and will gradually rise in an effort to approach the temperature at the wall followed by a gradual fall in response to cooling at the wall. In this situation, the lethality delivered by the process will be the result of the transient time–temperature history experienced by the product at the slowest-heating location in the can; this is usually the geometric center. Therefore, the ability to determine this time–temperature history accurately is of paramount importance in the calculation of thermal processes. In this section we review the various modes of heat transfer found in canned foods, and describe methods of temperature measurement and recording and how these data are treated for subsequent use in thermal process calculation.

### *5.4.2. Heat transfer modes*

Solid-packed foods in which there is essentially no product movement within the container, even when agitated, heat largely by conduction heat transfer. Because of the lack of product movement and the low thermal diffusivity of most foods, these products heat very slowly and exhibit a nonuniform temperature distribution during heating and cooling caused by the temperature gradient that is set up between the can wall and geometric center. For conduction-heating products, the geometric center is the slowest heating point in the container. Therefore, process calculations are based on the temperature history experienced by the product at the can center. Solid-packed foods such as canned fish and meats, baby foods, pet foods, pumpkin, and squash fall into this category. These foods are usually processed in still cook or continuous hydrostatic retorts that provide no mechanical agitation.

Thin-bodied liquid products packed in cans such as, milk, soups, sauces and gravies will heat by either natural or forced convection heat transfer, depending on the use of mechanical agitation during processing. In a still cook retort that provides no agitation, product movement will still occur within the container because of natural convective currents induced by density differences between the warmer liquid near the hot can wall and the cooler liquid near the can center (Datta and Teixeira, 1987, 1988). The rate of heat transfer in nearly all *convection-heating* products can be increased substantially by inducing forced convection through mechanical agitation. For this reason, most convection-heating foods are processed in agitating retorts designed to provide either axial or end-over-end can rotation. Normally, end-over-end rotation is preferred and can be provided in batch retorts, while continuous agitating retorts can provide only limited axial rotation.

Unlike conduction-heating products, because of product movement in forced convection-heating products, the temperature distribution throughout the product is reasonably uniform under mechanical agitation. In natural convection, the slowest heating point is somewhat below the geometric center and should be located experimentally in each new case (Figures 5.3a and 5.3b).

### *5.4.3. Heat penetration measurement*

The primary objective of heat penetration measurements is to obtain an accurate recording of the product temperature at the can cold spot over time while the container is being treated under a controlled set of retort processing conditions. This is normally accomplished through the use of copper–constantan thermocouples inserted through the can wall so as to have the junction located at the can geometric center. Thermocouple lead wires pass through a packing gland in

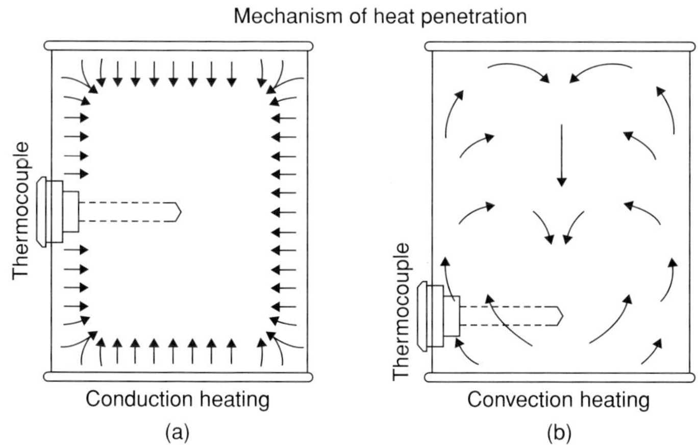


**Figure 5.3.** Conduction and convection heat transfer in canned foods of various composition from solid to liquid (a), and cold spot location in case of conduction and natural convection modes heat transfer (b).

the wall of the retort for connection to a data acquisition system in the case of a still cook retort. For agitating retorts, the thermocouple lead wires are connected to a rotating shaft for electrical signal pick up from the rotating armature outside the retort. Specially designed thermocouple fittings are commercially available for these purposes (Stumbo, 1965; NFPA, 1980; Lopez, 1987).

The precise temperature–time profile experienced by the product at the can center will depend on the physical and thermal properties of the product, size, and shape of the container, and retort operating conditions. Therefore, it is imperative that test cans of product used in heat penetration tests be truly representative of the commercial product with respect to ingredient formulation, fill weight, headspace, can size, and so on. In addition, the laboratory or pilot plant retort being used must accurately simulate the operating conditions that will be experienced by the product during commercial processing on the production-scale retort systems intended for the product. If this is not possible, heat penetration tests should be carried out using the actual production retort during scheduled breaks in production operations.

During a heat penetration test, both the retort temperature history and product temperature history at the can center are measured and

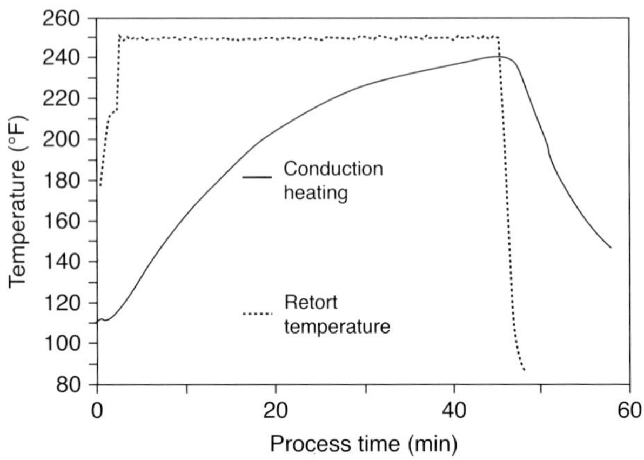


**Figure 5.4.** Generic heat penetration curve for a conduction-heating food during a thermal process.

recorded over time. A typical test process will include venting of the retort with live steam to remove all atmospheric air, then closing the vents to bring the retort up to operating pressure and temperature. This is the point at which "process time" begins, and the retort temperature is held constant over this period of time. At the end of the prescribed process time, the steam is shut off and cooling water is introduced under overriding air pressure to prevent a sudden pressure drop in the retort. This begins the cooling phase of the process, which ends when the retort pressure returns to atmosphere and the product temperature in the can has reached a safe low level for removal from the retort. A typical temperature–time plot of these data is shown in Figure 5.4, and illustrates the degree to which the product center temperature in the can lags behind the retort temperature during both heating and cooling.

### *5.4.4. Heat penetration curves and thermal diffusivity*

The response of the product temperature at the can center to the steam retort temperature applied at the can wall is governed by the physical laws of heat transfer and can be expressed mathematically. This mathematical expression is a model that serves as a basis for obtaining effective values for thermal properties of canned foods in order to use numerical computations on high-speed computers that are capable of simulating the heat transfer in thermal processing of canned foods.

A heat balance between the heat absorbed by the product and the heat transferred across the can wall from the steam retort could be expressed as follows for an element of food volume facing the can wall of surface area $A$ and thickness $L$:

$$\rho LAC_{\mathrm{p}}\frac{dT}{dT} = \frac{k}{L}A(T_{\mathrm{R}} - T) \tag{5.4}$$

where $T$ is product temperature, $T_{\mathrm{r}}$ is retort temperature, and $\rho$, $C_{\mathrm{p}}$, and $k$ are density, specific heat, and thermal conductivity of the product, respectively. Because of high surface heat transfer coefficient of condensing steam at the can wall and high thermal conductivity of the metal can, overall surface resistance to heat transfer can be assumed negligible, in contrast to the product's resistance to heat transfer. After rearranging terms, equation (5.4) can be written in the form of an ordinary differential equation:

$$\frac{dT}{dt} = \frac{k}{\rho C_{\mathrm{p}}}L^2(T_{\mathrm{R}} - T) \tag{5.5}$$

By letting the thermal diffusivity ($\alpha$) represent the combination of thermal and physical properties ($k/\rho C_{\mathrm{p}}$), and letting $T_{\mathrm{o}}$ represent the initial product temperature, the solution to equation (5.5) becomes:

$$\frac{T_{\mathrm{R}} - T}{T_{\mathrm{R}} - T_0} = \exp\left(\frac{\alpha}{L^2}t\right) \tag{5.6}$$

Thus, the product center temperature can be seen to be an exponential function of time; a semi-log plot of the temperature difference ($T_{\mathrm{R}} - T$) against time would produce a straight line sloping downward, having a slope related to the product's thermal diffusivity and can dimensions (Figure 5.5). The heat penetration rate factor ($f_{\mathrm{h}}$) is the reciprocal slope of the heat penetration curve (time required for one log cycle temperature change). Therefore, it can be related to the overall apparent thermal diffusivity of the product and container dimensions for a given container shape. For a finite cylinder the following relationship can be used to obtain the thermal diffusivity, $\alpha$,

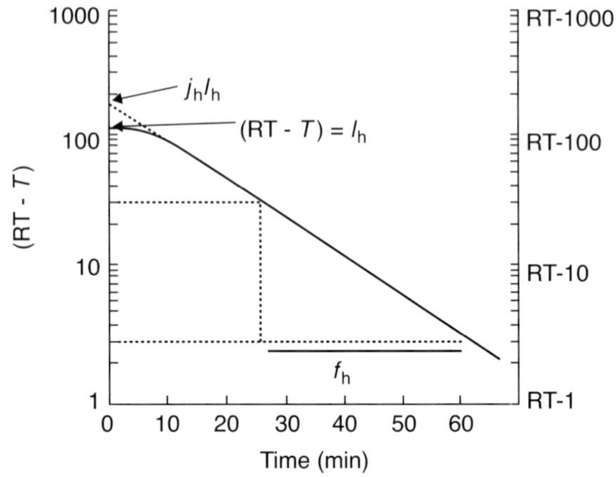


**Figure 5.5.** Semi-log heat penetration curve showing unaccomplished temperature difference (on log scale) versus time, from which heating rate ($f_h$) and heating lag ($j_h$) factors can be estimated.

from the heating rate factor taken from a heat penetration curve (Ball and Olson, 1957; Stumbo, 1965):

$$\alpha = \frac{0.398}{1/R^2 + (0.427/H^2)f_h} \tag{5.7}$$

where $R$ is the can radius in inches, $H$ is one-half the can height in inches, $f_h$ the heating curve slope factor in minutes, and $\alpha$ the product thermal diffusivity in compatible units. This relationship is also useful to determine the heating rate factor for the same product in a different size container, since thermal diffusivity is a combination of thermal and physical properties that characterize the product and its ingredient formulation, and remains unaffected by container size or shape. Similar relationships appropriate for other regular geometries can be found in published literature (Ball and Olson, 1957).

Another important heat penetration parameter obtained from the semi-log heat penetration curve is the heating lag factor ($j_{ch}$), which is taken as the ratio of the difference between the retort temperature ($T_R$) and pseudo-initial temperature ($T_o$), the temperature at which an extension of the straight-line portion of the heating curve intersects the ordinate axis ($T_R - T_o$) over the difference between retort temperature and actual initial product temperature ($T_R - T_i$). The heating lag factor can be used with deterministic conduction heat transfer models to account for heat transfer mechanisms other than pure conduction that often take place in most canned foods (Ball and Olson, 1957; Teixeira *et al*., 1999).

## 5.5. Process design

Once a heat penetration curve has been obtained from laboratory heat penetration data or predicted by a computer model, there are essentially two widely accepted methods for using these data to perform thermal process calculations for determining process time at retort temperature (process design). The first of these is the general method of process calculation (Bigelow *et al*., 1920), and the second is the Ball formula method of process calculation (Ball and Olson, 1957). Only the general method is described in this chapter because of its connected use with mathematical heat transfer models.

As the name implies, the general method is the most versatile method of process calculation because it is universally applicable to essentially any type of thermal processing situation. It makes direct use of the product temperature history at the can center obtained from a heat penetration test (or predicted by a mathematical model) to evaluate the integral shown in equation (5.3) for calculating the process lethality delivered by a given temperature–time history. A straightforward numerical integration of equation (5.3) can be expressed as follows with reference to Figure 5.6:

$$F_o = \sum_{i=1}^{n} \Delta F_i = \sum_{i=1}^{n} 10^{[(T_i-121)/Z]\Delta t} \tag{5.8}$$

Figure 5.6 is a direct plot of the can center temperature experienced during a heat penetration test. Since no appreciable lethality can occur until the product temperature has reached the lethal temperature range (above 105°C), equation (5.8) need only be evaluated over the time period during which the product temperature remains above 105°C. By dividing this time period into small time intervals ($\Delta t$) of short duration as shown in Figure 5.6. The temperature $T_i$ at each

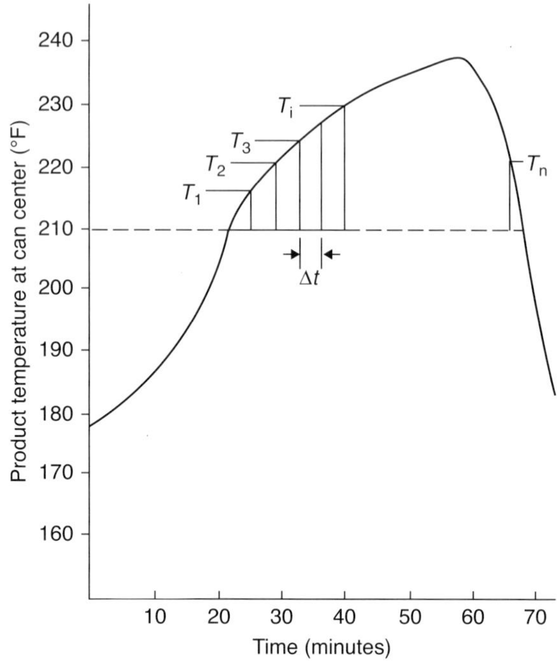


**Figure 5.6.** Temperature history at center of canned food during thermal process for calculation of process lethality by general method.

time interval can be read from the curve and used to calculate the incremental lethality ($\Delta F_i$) accomplished during that time interval. Then the sum of all these incremental sterilizing values equals the total lethality, $F_o$, delivered by the test process. To determine the process time required to deliver a specified lethality, the cooling portion of the curve in Figure 5.6 is shifted to the right or left and the integration is repeated until the delivered *lethality* so calculated agrees with the value specified for the process.

When first introduced in 1920, this method was sometimes referred to as the graphical trial-and-error method because the integration was performed on specially designed graph paper to ease the tedious calculations that were required. The method was also time consuming, and soon gave way in popularity to the historically more convenient (but less accurate) Ball formula method. With the current widespread availability of low-cost programmable calculators and desktop computers, these limitations are no longer of any consequence, and the general method is currently the method of choice because of its accuracy and versatility.

The general method is particularly useful in taking maximum advantage of computer-based data logging systems used in connection with heat penetration tests. Such systems are capable of reading temperature signals received directly from thermocouples monitoring both retort temperature and product center temperature, and processing these signals through the computer. Through programming instructions, both retort temperature and product center temperature are plotted against time without any data transformation. This allows the operator to see what has actually happened throughout the duration of the process test. As the data are being read by the computer, additional programming instructions call for calculation of the incremental process lethality ($\Delta F_i$) at each time interval between temperature readings and summing these over time as the process is under way. As a result, the accumulated lethality ($F$) is known at any time during the process and can be plotted on the graph along with the temperature histories to show the final value reached at the end of the process.

Another test can be repeated for a longer or shorter process time with instant results on the $F_o$ achieved. By examining the results from both tests, the desired process time for the target $F$-value can be closely estimated and then quickly tested for confirmation. The results of two such heat penetration tests are shown superimposed on each other in Figure 5.7. These results show that test 1, with a process time of 68 min, produced an $F$-value of 6, and test 2, with a process time of 80 minutes, produced an $F$-value of 8, suggesting that a target $F$-value of 7 will be achieved by an intermediate process time. This can be confirmed by running a test at the suggested process time, and examining the resulting $F$-value.

## 5.6. Mathematical model for heat transfer

The development and use of mathematical heat transfer models for simulating the thermal processing of canned foods has been well documented in published scientific literature (Teixeira *et al.*, 1969,

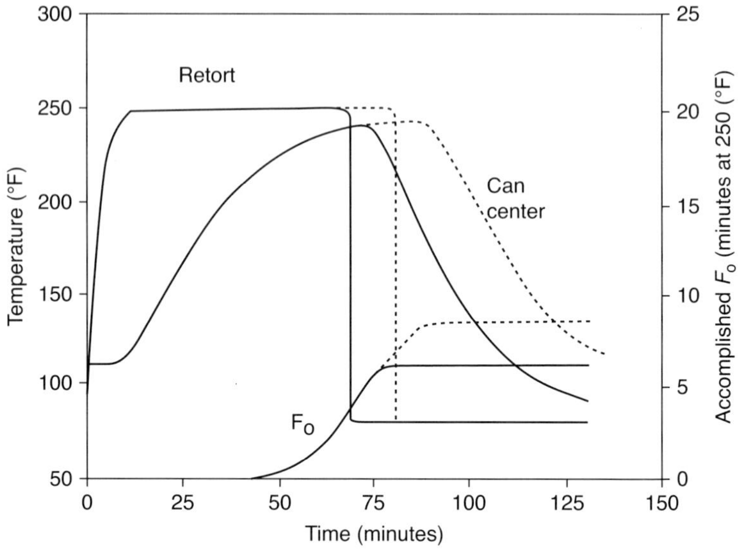


**Figure 5.7.** Computer-generated plot of retort temperature, can center temperature, and accomplished lethality ($F_o$) over time for two different process times superimposed on each other.

1975; Teixeira and Manson, 1982; Datta *et al.*, 1986). The model described in this chapter makes use of a numerical solution by finite differences of the two-dimensional partial differential equation that describes conduction heat transfer in a finite cylinder. During conduction heating, heat is applied only at the can surface, temperatures will rise first only in regions near the can walls, while temperature near the can center will begin to respond only after a considerable time lag. Mathematically, the temperature is a distributed parameter in that at any point in time during heating, the temperature takes on a different value with location in the can; and in any one location, the temperature changes with time as heat gradually penetrates the product from the can walls toward the center.

The mathematical expression that describes this temperature distribution pattern over time is shown in Figure 5.8 and lies at the heart of the numerical computer model. This expression is the classic second order partial differential equation for two-dimensional unsteady heat conduction in a finite cylinder, and can be written in the form of

$$\frac{\partial T}{\partial t} = \alpha\left[\frac{\partial^2 T}{\partial r^2} + \frac{1}{r}\frac{\partial T}{\partial r} + \frac{\partial^2 T}{\partial h^2}\right]$$

where
$T$ = temperature
$t$ = time
$\alpha$ = thermal diffusivity
$r$ = radial position in cylinder
$h$ = vertical position in cylinder

**Figure 5.8.** Two-dimensional second order partial differential equation for conduction heat transfer (heat conduction equation) in a finite cylinder. (From Teixeira and Manson (1982).)

finite differences for numerical solution by digital computer, as shown in Figure 5.9. The finite differences are discrete increments of time and space defined as small intervals of process time and small increments of container height and radius ($\Delta t$, $\Delta h$, and $\Delta r$, respectively).

As a framework for computer iterations, the cylindrical container is imagined to be subdivided into volume elements that appear as layers of concentric rings having rectangular cross-sections, as illustrated in

$$T_{(ij)}^{(t+\Delta t)} = T_{(ij)}^{(t)} + \frac{\alpha\Delta t}{\Delta r^2}\left[T_{(i-1,j)} - 2T_{(ij)} + T_{(i+1,j)}\right]^{(t)} + \frac{\alpha\Delta t}{2r\Delta r}\left[T_{(i-1,j)} - T_{(i+1,j)}\right]^{(t)} + \frac{\alpha\Delta t}{\Delta h^2}\left[T_{(i,j-1)} - 2T_{(i,j)} + T_{(i,j+1)}\right]^{(t)}$$

where $\Delta t$, $\Delta r$, $\Delta h$ = discrete increments of time, radius, and height, and i and j denote sequence of radial and vertical increments away from can wall and mid-plane.

**Figure 5.9.** Heat conduction equation for finite cylinder expressed in finite differences for numerical solution by computer iteration. (From Teixeira and Manson (1982).)

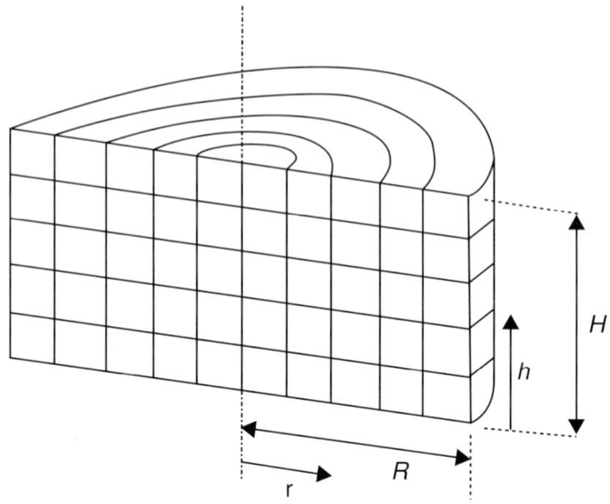


**Figure 5.10.** Subdivision of a cylindrical container for application of finite differences. (From Teixeira *et al*. (1969).)

Figure 5.10 for the upper half of the container. Temperature nodes are assigned at the corners of each volume element on a vertical plane, as shown in Figure 5.11, where $I$ and $J$ are used to denote the sequence of radial and vertical volume elements, respectively. By assigning appropriate boundary and initial conditions to all the temperature nodes (interior nodes set at initial product temperature, and surface nodes set at retort temperature), the new temperature reached at each node can be calculated after a short time interval ($\Delta t$) that would be consistent with the thermal diffusivity of the product obtained from heat penetration data ($f_{\mathrm{h}}$). This new temperature distribution is then taken to replace the initial one, and the procedure repeated to calculate the temperature distribution after another time interval. In this way, the temperature at any point in the container at any instant in time is obtained. At the end of process time, when steam is shut off and cooling water is admitted to the retort, the cooling process is simulated by simply changing the boundary conditions from retort temperature $T_{\mathrm{R}}$ to cooling temperature $T_{\mathrm{c}}$ at the surface nodes and continuing with the computer iterations described above.

The temperature at the can center can be calculated after each time interval to produce a predicted heat penetration curve upon which the process lethality, $F$, can be calculated. When the numerical computer

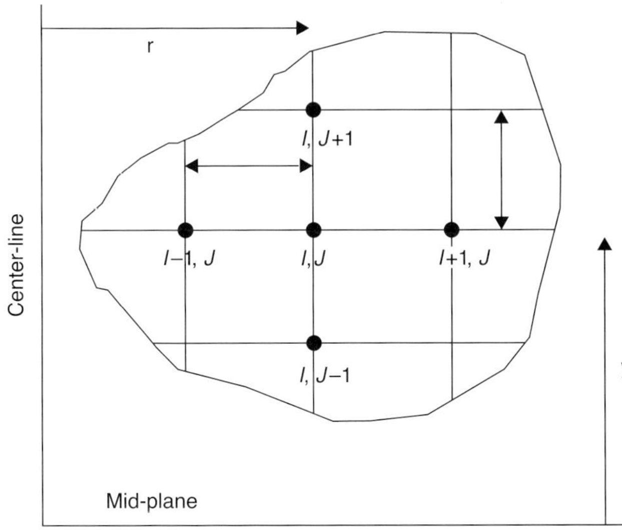


**Figure 5.11.** Labeling of grid nodes in matrix of volume elements on a vertical plan for application of finite differences. (From Teixeira *et al.* (1969).)

model is used to calculate the process time required at a given retort temperature to achieve a specified lethality, $F$, the computer follows a programmed search routine of assumed process times that quickly converges on the precise time at which cooling should begin in order to achieve the specified $F$-value. Thus, the model can be used to determine the process time required for any given set of constant or variable retort temperature conditions.

## 5.7. Process deviations

Control of thermal process operations in food canning factories has consisted of maintaining specified operating conditions that have been predetermined from product and process heat penetration tests, such as the process calculations for the time and temperature of a batch cook. Sometimes unexpected changes can occur during the course of the process operation or at some point upstream in a processing sequence such that the prespecified processing conditions are no longer valid or appropriate, and off-specification product is produced that must be either reprocessed or destroyed at appreciable economic loss. These types of situations are known as process deviations, and can be of critical importance in food processing operations because the physical process variables that can be measured and controlled are often only indicators of complex biochemical reactions that take place under the specified process conditions.

Because of the important emphasis placed on the public safety of canned foods, processors operate in strict compliance with the Food and Drug Administration (FDA) Low-Acid Canned Food (LACF) regulations. Among other things, these regulations require strict documentation and record-keeping of all critical control points in the processing of each retort load or batch of canned product. Particular emphasis is placed on product batches that experience an unscheduled process deviation, such as when a drop in retort temperature occurs during the course of the process that may result from loss of steam pressure. In such a case, the product will not have received the established scheduled process, and must be either destroyed, fully reprocessed or set aside for evaluation by a competent process authority. If the product is judged to be safe, then batch records must contain documentation showing how that judgment was reached. If judged unsafe, then the product must be fully reprocessed or destroyed. Such practices are costly.

In recent years, food engineers knowledgeable in the use of engineering mathematics and scientific principles of heat transfer have developed computer models capable of simulating thermal processing of conduction-heated canned foods as those described in this chapter. These models make use of numerical solutions to mathematical heat transfer equations capable of predicting accurately the internal product cold spot temperature in response to any dynamic temperature experienced by the retort during the process. As such, they are very useful in the rapid evaluation of deviations that may unexpectedly occur.

Accuracy of such models is of paramount importance, and the models must work equally as well for any mode of heat transfer or size and shape container. Recall that the deterministic model described earlier in this chapter was derived for the case of pure conduction heat transfer in a solid body of finite cylinder shape. It would not be applicable to the many food products that heat by convection or to varying degrees of combined convection and conduction, nor to different shapes. Recent work reported in the literature has described effective modification and simplification of the model to overcome these limitations (Noronha *et al.*, 1995; Teixeira *et al.*, 1999). These reports confirmed that food containers need not be of the same shape as the solid body assumed by the heat transfer model. They could be of any shape so long as temperature predictions were required only at the single cold spot location within the container from which heat penetration data were determined.

The improved model assumed the product was a pure conduction-heating solid in the form of a sphere. An "apparent" thermal diffusivity was obtained for the solid sphere that would produce the same heating rate as that experienced by the product cold spot. Similarly, the precise radial location where the heating lag factor ($j_h$) was the same as that at the product cold spot, would be used as the location at which temperature is calculated by the model (Figures 5.12 and 5.13). Thus, for any product with empirical parameters ($f_h$ and $j_h$) known from heat penetration tests, it would be possible to simulate the thermal response at the product cold spot to any dynamic boundary condition (time varying retort temperature) regardless of container size or shape or process conditions (mode of heat transfer).

Recall that heat penetration test data normally produce straight line semi-log heat penetration curves from which the empirical heat penetration parameters ($f_h$ and $j_h$) can be determined. Incorporation of the parameters into the heat transfer model is accomplished by the relationship between thermal diffusivity ($\alpha$) and heating rate factor ($f_h$) for a sphere (equation (5.9)); and the relationship between heating lag factor ($j_h$) and radial location ($r$) within the sphere (equation (5.10)). These and similar relationhips for other regular solid body shapes can be found in the literature (Ball and Olson, 1957; Noronha *et al.*, 1995).

$$f_h = 0.233(R^2/\alpha) \tag{5.9}$$

$$j(r) = 0.637(R^2/r)\sin(\pi r/R) \tag{5.10}$$

Results from heat penetration tests on five products (Teixeira *et al.*, 1999) are presented in Table 5.4. All products exhibited straight line

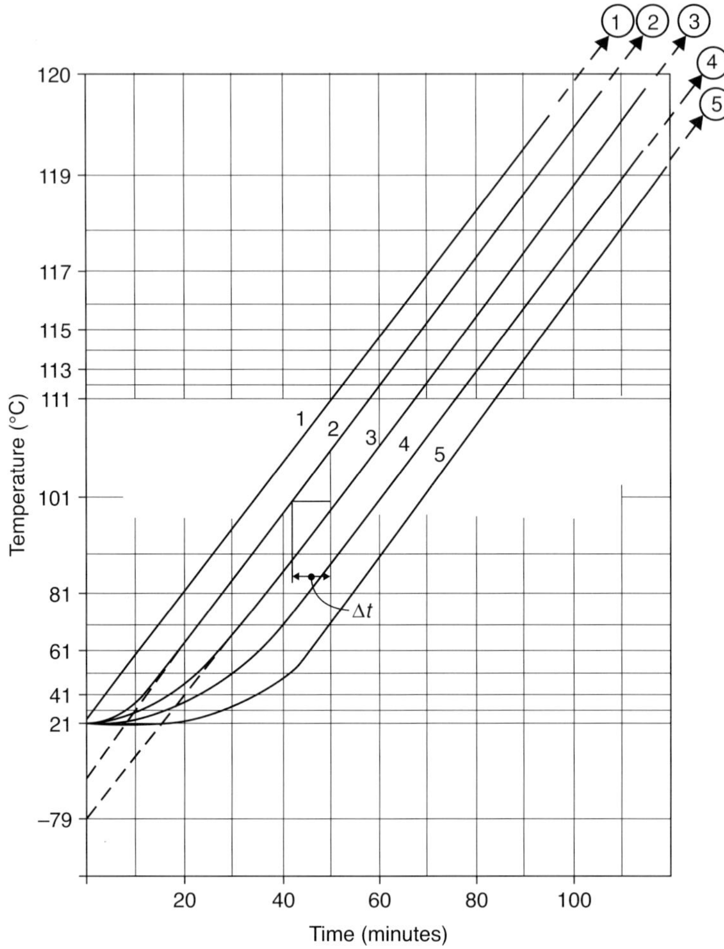


**Figure 5.12.** Heat penetration curves for five different locations along the radius on the midplane of a cylindrical container (see Figure 5.15), illustrating relationship between location and heating lag factor ($j_h$).

(log-linear) heat penetration curves on semi-log plots of unaccomplished temperature differences versus time. Can-to-can variation in heating rate factor ($f_h$) and lag factors derived from direct analysis of the heat penetration curve ($j_h$ analyze) were determined by the maximum and minimum values found over all six cans from two replicate tests. The true heating lag factor found by trial and error

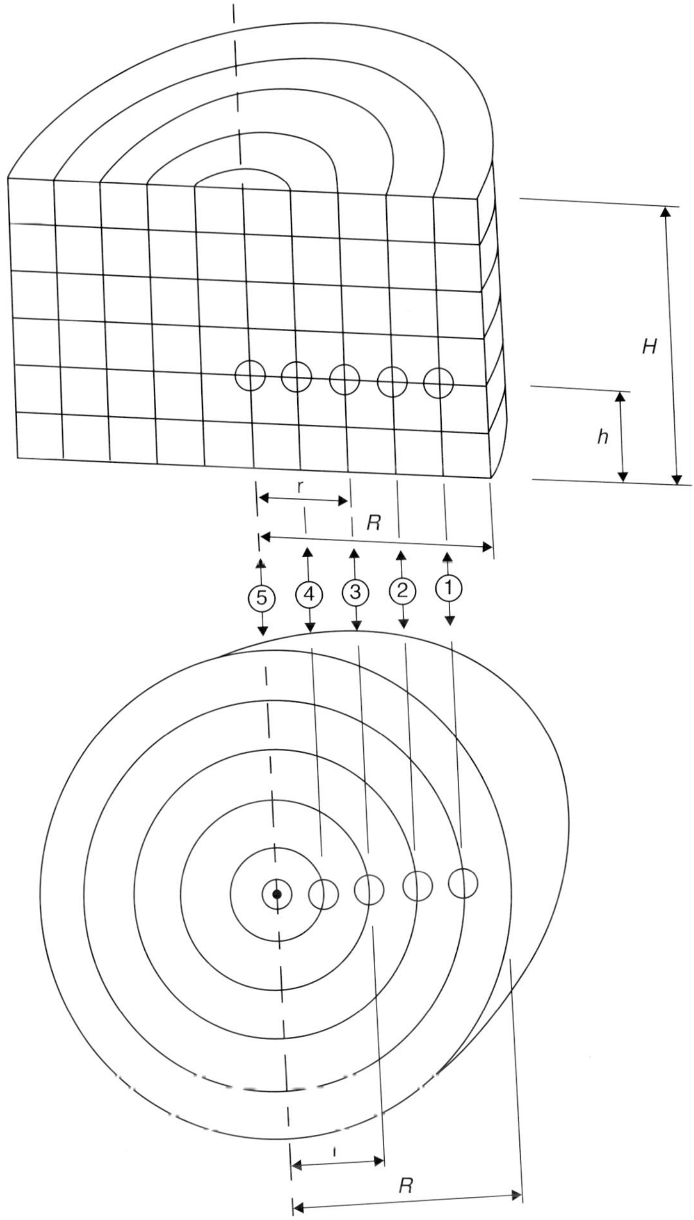


**Figure 5.13.** Replacement of solid body shape from finite cylinder to perfect sphere for simplification of numerical heat transfer model with choice of radial location based upon heating lag factor from heat penetration tests.

**Table 5.4.** Heat penetration results on products using two replicated heat penetration tests with six instrumented cans for each product

| Product and process | $f_h$ (minutes) (range) | $j_h$ analyze (range) | $j_h$ simulate | $F_o$ actual (range) | $F_o$ simulated |
|---|---|---|---|---|---|
| 5% Bentonite 1 kg cans (98 × 110 mm); static cook | 70.4–73.0 | 1.9–2.0 | 2.0 | 6.0–7.0 | 6.2 |
| 5% Bentonite tuna cans (86 × 45 mm); static cook | 20.0–22.0 | 1.4–1.6 | 1.4 | 7.5–9.8 | 7.4 |
| Water 1 kg cans (98 × 110 mm); static cook | 3.0–3.1 | 1.8–2.3 | 1.0 | 9.8–10.8 | 9.9 |
| Water tuna cans (86 × 45 mm); static cook | 1.7–1.9 | 2.5–3.9 | 1.0 | 7.9–10.6 | 7.7 |
| Peas in brine half-kg cans (74 × 88 mm); agitated cook | 2.5–3.0 | 2.6–3.4 | 1.0 | 10.8–12.0 | 11.0 |

*Source*: Teixeira *et al*., 1999.

simulation ($j_h$ simulate) was also compared. This was the value chosen for use in the heat transfer model along with the maximum $f_h$ values (slowest heating) for conservative routine simulation of each product. The range of lethality values calculated from the temperatures measured by thermocouples in each can ($F_o$ actual) were also compared. Lethality was calculated from the simulated temperature profile ($F_o$ simulated) predicted by the heat transfer model in response to the retort temperature data file from each heat penetration

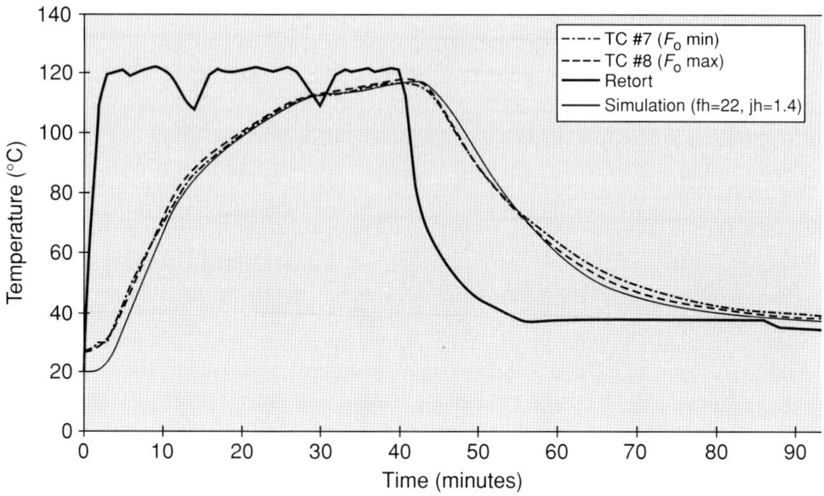


**Figure 5.14.** Comparison of internal cold spot temperatures predicted by model simulation with those measured by thermocouples in response to multiple retort temperature deviations during a heat penetration test with 5% bentonite suspension in 6-ounce tuna can. (From Teixeira *et al.* (1999).)

test as input. Figure 5.14 compares internal cold spot temperatures predicted by model simulation with profiles measured by thermocouple in response to multiple retort temperature deviations during a heat penetration test (Teixeira *et al.*, 1999). The simulated profiles follow the measured profiles quite closely in response to relatively severe and twice repeated deviations.

The final test of model performance in the simulation and evaluation of process deviations was a comparison of lethalities accomplished by actual and simulated temperature profiles (Table 5.5). Recall that the accomplished lethality ($F_0$) for any thermal process is easily calculated by numerical integration of the measured or predicted cold spot temperature over time as explained previously. Thus, if the cold spot temperature can be accurately predicted over time, so can accumulated process lethality. In all cases the simulated lethality predicted agreed most closely with the minimum actual lethality calculated from measured temperature profiles. Model predictions that tend toward the minimum side of the range are always desirable for conservative decision making.

**Table 5.5.** Process deviation test results showing lethalities calculated from temperatures predicted by model simulation ($F_0$ simulated), and those calculated from actual measured temperatures ($F_0$ actual) in slowest and fastest heating cans of the same product in response to different types of retort temperature deviations during processing

| Product/process | Deviation Type (A, B, C) | $F_0$ simulated | $F_0$ actual | |
|---|---|---|---|---|
| | | | Min | Max |
| 5% Bentonite 1 kg | A | 5.5 | 5.5 | 6.4 |
| | B | 3.0 | 3.3 | 5.3 |
| static | C | 1.7 | 1.6 | 2.4 |
| 5% Bentonite tuna | A | 6.5 | 6.6 | 7.6 |
| | B | 5.6 | 5.7 | 7.0 |
| static | C | 4.8 | 4.7 | 5.7 |
| Water 1 kg | A | 7.4 | 7.4 | 8.2 |
| | B | 7.8 | 8.8 | 10.3 |
| static | C | 7.1 | 7.4 | 8.8 |
| Water tuna | A | 4.4 | 5.4 | 6.2 |
| | B | 6.0 | 6.6 | 7.7 |
| static | C | 5.5 | 6.7 | 8.0 |
| Peas in Brine | C | 9.1 | 9.2 | 10.0 |
| Half-kg Agitated | | | | |

*Source*: Teixeira *et al*., 1999.

## 5.8. On-line correction of process deviations

The work described in the previous section was the first part of a two-phase project for developing and demonstrating a computer-based intelligent on-line control system for the operation of batch retorts for canned food sterilization. Key to this system was the performance of commercially available Can-Calc© thermal process simulation software (Engineering Cyber Solutions, Gainesville, FL). This software makes use of a mathematical heat transfer model similar to that described in this chapter. The model is capable of predicting the internal product cold spot temperature in response to any dynamic temperature of the retort. The accomplished lethality ($F_0$) for any thermal process can be easily calculated by numerical integration of the cold spot temperature profile over time using the General Method. Thus, if the cold spot temperature can be accurately predicted, so can the accumulated process lethality.

Using only the monitored retort temperature as input, the model can operate as a subroutine calculating the internal product cold spot temperature at small time intervals. At the same time, the model could calculate the process lethality associated with the cold spot temperature in real time while the process was under way. At each time step, the model would simulate additional lethality contributed by the cooling phase if cooling were to begin at that time. Thus, the decision of when to end heating and begin cooling would be withheld until the model had determined that final target process lethality would be reached at the end of cooling.

By programming the control logic to continue heating until the accumulated lethality has reached some designated target value, the process will always end with the desired level of lethality ($F_o$) regardless of an unscheduled process temperature deviation. At the end of the process, complete documentation of measured retort temperature history, calculated center temperature history, and accomplished lethality ($F_0$) can be generated in compliance with regulatory record-keeping requirements. Such documents are shown in Figure 5.15 for a normal process (above), and for the same intended process with an unexpected deviation (below) (Datta *et al.*, 1986).

## 5.9. Conclusions

This chapter has focused on the development and application of heat transfer models for simulation of thermal processing in the process development and manufacture of heat-sterilized canned foods. The models are capable of accurately predicting internal product temperature over time during heating in response to dynamic retort operating conditions for any degree of combined conduction–convection mode of heat transfer, and for any size or shape container.

Applications of these models through commercially available Can-Calc© software to process design, rapid off-line evaluation of process deviations, and real-time on-line computer control of retort operations was described in detail.

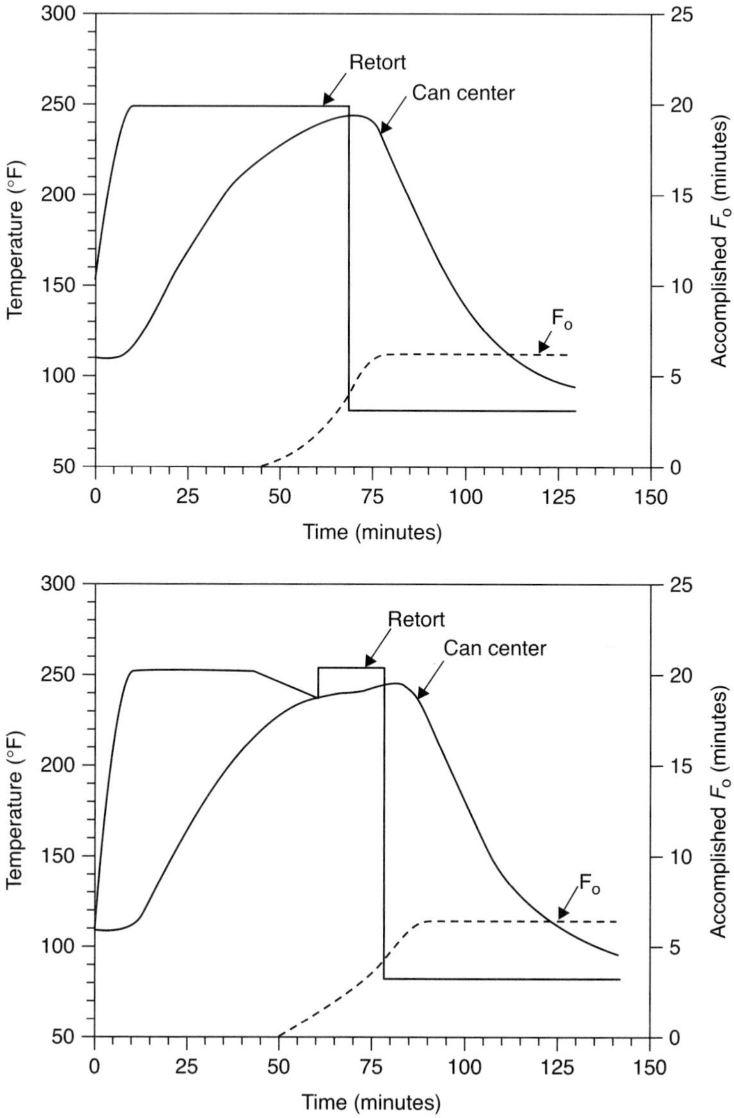


**Figure 5.15.** Computer-generated output from computer-based on-line control system showing scheduled heating time of 68 minutes for normal process (above), and heating time extended automatically to 76 minutes in compensation for unscheduled temporary loss of retort temperature (process deviation).

The chapter also presented appropriate review of underlying principles and concepts of thermal processing important to understanding model development, applications, and limitations. These included as follows:

1. Review of thermal death-time relationships, which describe how thermal inactivation of bacterial spore populations can be quantified as a function of time and temperature.
2. Process lethality and sterilizing value, defining concepts for specifying process requirements with respect to public health considerations and spoilage probability.
3. Heat transfer considerations, describing methods of temperature measurement and recording, and how these data are treated to obtain important heat penetration parameters for subsequent use in thermal process calculation and in deterministic models.
4. Process calculations, which described the general method for calculating thermal processes, including the process lethality delivered by a specific process, as well as the process time required at a given temperature to deliver a specified lethality value.
5. Process deviations, describing application to the rapid evaluation of unexpected process deviations.
6. On-line computer control, showing use of models for of unexpected process deviations.

## List of Symbols

| | |
|---|---|
| $A$ | Area through which heat transfer occurs, $m^2$ |
| $a$ | Initial number of viable bacterial spores at the beginning of a thermal process |
| $b$ | Final number of viable bacterial spores (survivors) at the end of a thermal process |
| $C$ | Concentration of primary component in a first-order reaction, quantity per unit mass or volume (e.g., spores/mL) |
| $C_p$ | Specific heat or heat capacity, kJ/kg °C |
| $D$ | Decimal reduction time, time for one log cycle reduction in population during exposure to constant lethal temperature, min |

| | |
|---|---|
| $D_r$ | Decimal reduction time at a specified reference temperature ($T_r$), minute |
| $D_{121}$ | Decimal reduction time at temperature of 121°C (250°F), minute |
| $F$ | Process lethality, minutes at any specified temperature and z-value, applied to destruction of microorganisms |
| $F_o$ | Lethality value applied to destruction of microorganisms with $z$-value of 10°C (18°F), min at 121.1°C (250°F) |
| $f_h$ | Heat penetration factor, time for straight-line portion of semi-log heat penetration curve to traverse one log cycle, minute |
| $H$ | Half-height of cylindrical food can, m |
| $h$ | Any distance along vertical dimension (height) from the mid-plane in a cylindrical can, m |
| $I$ | Sequence of radial nodal points for numerical iteration |
| $J$ | Sequence of vertical nodal points for numerical iteration |
| $j_h$ | Heating lag factor at geometric center of food container, dimensionless |
| $k$ | Thermal conductivity, W/m k, with reference to heat transfer |
| $L$ | Length or thickness, m |
| $R$ | Radius of cylinder or sphere, m |
| $r$ | Any distance along radius from centerline, m |
| $T$ | Temperature, °C |
| $T_c$ | Temperature of cooling medium, °C |
| $T_i$ | Initial product temperature, °C |
| $T_o$ | Pseudo-initial temperature, the temperature at which an extension of the straight line portion of the heat penetration curve intersects the ordinate axis, °C |
| $T_R$ | Retort temperature, °C |
| $T_r$ | Reference temperature at which $D_r$ is measured |
| $T_{(ij)}^{(t)}$ | Temperature at any grid node ($ij$) at time ($t$), °C |
| $T_{(ij)}^{(t+\Delta t)}$ | Temperature at any grid node ($ij$) at time ($t + \Delta t$), or one time interval ($\Delta t$) later,°C |
| $t$ | Time, minute |
| $x$ | Spatial location within a food container, m |

$Z$ Temperature-dependency factor in thermal inactivation kinetics, temperature difference required for tenfold change in decimal reduction time ($D$-value), °C

$\alpha$ Thermal diffusivity, $m^2/s$

$\Delta F_i$ Incremental lethality accomplished over time interval ($\Delta t$), minute

$\Delta h$ Vertical height of incremental volume element ring in finite difference solution to heat transfer equation, m

$\Delta r$ Radial width of incremental volume element ring in finite difference solution to heat transfer equation, m

$\Delta t$ Time interval between computational iterations in finite difference solution to heat transfer equation, minute

$\rho$ Product density, $kg/m^3$

## References

Ball, C.O. and Olson, F.C.W. 1957. *Sterilization in Food Technology*. McGraw-Hill, New York.

Bigelow, W.D., Bohart, G.S., Richardson, A.C., and Ball, C.O. 1920. Heat penetration in processing of canned foods. *National Canners Association Bulletin*, 16L.

Datta, A.K. and Teixeira, A.A. 1987. Numerical modeling of natural convection heating in canned foods. *Transactions of the ASAE*, 30(5): 1542–1551.

Datta, A.K. and Teixeira, A.A. 1988. Numerically predicted transient temperature and velocity profiles during natural convection heating of canned liquid foods. *Journal of Food Science*, 53(1): 191–196.

Datta, A.K., Teixeira, A.A., and Manson, J.E. 1986. Computer-based retort control logic for on-line correction of process deviations. *Journal of Food Science*, 51(2): 480–483, 507.

Lopez, A.A. 1987. *Complete Course in Canning, Book 1, Basic Information on Canning*, 11th ed. The Canning Trade, Baltimore, MD.

NFPA, 1980. *Laboratory Manual for Food Canners and Processors*, Vol. 1. AVI, Westport, CT.

Noronha, J., Hendrickx, M., Van Loey, A., and Tobback, P. 1995. New semi-empirical approach to handle time-variable boundary conditions during sterilization of non-conductive heating foods. *Journal of Food Engineering*, 24: 249–268.

Stumbo, C.R. 1965. *Thermobacteriology in Food Processing*. Academic Press, New York.

Teixeira, A.A., Balaban, M.O., Germer, S.P.M., Sadahira, M.S., Teixeira-Neto, R.O., and Vitali, A. 1999. Heat transfer model performance in simulation of process deviations. *Journal of Food Science*, 64(3): 488–493.

Teixeira, A.A., Dixon, J.R., Zahradnik, J.W., and Zinsmeister, G.E. 1969. Computer optimization of nutrient retention in thermal processing of conduction-heated foods. *Food Technology*, 23(6): 137–143.

Teixeira, A.A., Zinsmeister, G.E., and Zahradnik, J.W. 1975. Computer simulation of variable retort control and container geometry as a possible means of improving thiamine retention in thermally processed foods. *Journal of Food Science*, 40: 656–662.

Teixeira, A.A. and Manson, J.E. 1982. Computer control of batch retort operations with on-line correction of process deviations. *Food Technology*, 36(4): 85–90.

# Chapter 6

# OPTIMIZATION, CONTROL, AND VALIDATION OF THERMAL PROCESSES FOR SHELF-STABLE PRODUCTS

*François Zuber, Antoine Cazier, and Jean Larousse*

## 6.1. Introduction

The need for industrial guidelines in the design and validation of thermal treatments for shelf-stable products is obvious, according to the very serious food safety implications in case of misunderstanding of technical details or lack of know-how, while taking into account the multiple critical points to be checked, during thermal processing of foods.

The regulation (both in Europe and the United States) is detailed and clear about the specifications of thermal processing of foods to render them shelf-stable—commercial sterility assumed by no microbiological growth after incubations of samples—but all complex factors, leading to proper design, thermal process development, and their use, may not always be integrated by food processors.

Therefore, a complete HACCP procedure was conducted by French canners, with the support of their Technical Center, to write and edit an exhaustive Guideline of Good Practices, intended to help processors check all critical points. The procedure for the validation of a newly designed thermal treatment was described as a technical

*Thermal Processing of Foods: Control and Automation*. Edited by K.P. Sandeep

reference guideline, from the initial calculation of scheduled process to prototype series of stabilized food, leading to the rules to be observed before going into production steps. This guideline was written both for packaged retorted foods, and continuously bulk treated product (associated with aseptic packaging). This chapter gives a translation of all main points of this Guideline of Good Practices.

Application of the general procedures used when validating thermal processes in a retort can be illustrated by the well-known "Guideline for the heat distribution study in batch retorts" designed and edited by the FDA (these guidelines were initially described for water cascading retorts, but can be used for other heating media too). When performing the heat distribution studies in several canneries in Europe, mainly in France, we introduced the use of total $F_0$ achieved in the heating medium during complete sterilization cycle (heating followed by cooling) as an additional indicator called "Sterilization Power Integrated Intensity," for the evaluation of proper heat distribution and evaluation of the position of "critical zone" in the retort. The practical use of this indicator is illustrated with an example taken from actual measurements in a water cascading retort.

The necessary target for absolute food safety of these foods also meets the need for a control of cooking effects (nutritional and organoleptic properties). The main tool used for the design of shelf-stable low-acid foods is the sterilization value (and/or pasteurization value for acid foods or reduced $a_W$ foods). Applying the properly designed thermal treatment, expected to ensure target sterilization value at the product cold point, leads to shelf stability. This requires predictive calculation of time–temperature treatments, based both on initial microbiological load and thermal transfer properties of the food, and also the characteristics of the thermal processing equipment.

The main improvement developed during the past 20 years, in designing thermal processes, was the integration of the cooling phase in the total treatment, as an important means of the sterilization during process. Taking into account the cooling phase of retorting is not always done, or only partially done when using the Ball method for the calculation of scheduled processes.

In Northern Europe, the use of sterilization value achieved at cold point of the product, *at the end of the heating phase*, as a target for required intensity of thermal treatment, has been the practical rule for years. This leads very often to overcooking of products, associated with unnecessary oversterilization, especially when more than half of the total sterilization value at the critical point is achieved during the cooling phase of products. However, another approach was used in southern European countries (i.e., Spain, Italy, France) with more interest in the preservation of organoleptic properties of foods, while keeping the food safety as first target.

In most cases, the classical Ball method calculation of thermal treatments leads to thermal processes that are very suitable for packaged foods with convective heating behavior. However, the Ball method underestimates the inertial heating effect in conduction heating products during the beginning of the cooling phase of retort ("overshooting" of temperature at the product cold point). This often leads to extensive thermal processes, with an actual $F_0$ achieved at the critical point being 30–50% higher than the target $F_0$ value, and poor retention of the organoleptic and nutritional values of the food.

In order to allow proper optimization of thermal treatments, even in the case of strictly conduction heating products in large packages, a specially designed software was developed with the help of academic institutions, to minimize the total thermal treatment of batch retorted foods: the duration of the heating phase is automatically calculated through a step-by-step iterative calculation, using a simulation of the forthcoming "end of heating phase" and simulating the forthcoming temperature profile in the product during the cooling phase in the retort. The structure and performance of this optimization software is illustrated below.

## 6.2. Regulatory considerations

Due to serious food safety implications in case of spoilage or non-stability, the manufacturing of shelf stable, thermally stabilized food products (usually called "canned food" or "retorted food") is associated with complex technical and scientific backgrounds and precise regulations in the different countries. Various interpretations of food safety assessments and means of control and validation (thermal processing at a predetermined $F_0$ value, incubation of samples, stability assessment, etc.) exist, in accordance with different regulations.

A retorted, or canned, food can be defined as follows.
*According to Codex (1993) definitions:*

"Canned food" means commercially sterile food in hermetically sealed containers.
"Commercial sterility of thermally processed food" means the condition achieved by application of heat, sufficient, alone or in combination with other appropriate treatments, to render the food free of microorganisms capable of growing in the food at normal nonrefrigerated conditions at which the food is likely to be held during distribution and storage.

*According to the US definition:*

*21 Code of Federal Regulations, part 113: Thermally processed low-acid food packaged in hermetically sealed containers*
"Low-acid food" means any food other than alcoholic beverages, with a finished equilibrium pH greater than 4.6 and a water activity greater than 0.85.
"Commercial sterility" of thermally processed (low-acid) food means the condition achieved:

1. By the application of heat that renders the food free of:
    - microorganisms capable of reproducing in the food under nonrefrigerated conditions of storage and distribution, and
    - viable microorganisms (including spores) of public health significance.

Or:

2 By the control of water activity and the application of heat that renders the food free of microorganisms capable of reproducing in the food under nonrefrigerated conditions of storage and distribution.

*9 Code of Federal Regulations, subpart 318-300: Canning and canned [meat] products*
"Canned product" means a meat food product with a water activity above 0.85, which receives a thermal process either before or after being packed in a hermetically sealed container.

"Shelf stability (or commercial sterility)" means the condition achieved by application of heat, sufficient, alone or in combination with other ingredients and/or treatments, to render the product free of microorganisms capable of growing in the product at nonrefrigerated conditions (over 10°C) at which the product is intended to be held during distribution and storage.

*According to French regulation (JORF, 1955):*

Canned foods are perishable foods from vegetal or animal origin, for which the preservation (at room temperature) is obtained by the combined use of the following techniques:

- Packaging in a liquid-tight and airtight sealed container, avoiding any microorganism penetration, for all temperatures up to 55°C.
- Treatment with heat, or any other authorized treatment, in order to ensure destruction or total inhibition of both enzymes and microorganisms and their toxins, for which the presence or proliferation could spoil the food or make it unsuitable for human feeding.

Starting from the definition of the nature of canned foods, the minimal requirements for the characteristics of the final products and sometimes for good manufacturing practices, are also defined in the regulations. There are differences in the ways regulations are formulated in different countries, through all have the same goal—ensuring the safety of shelf-stable retorted foods. Some regulations are formulated in the following manner:

↳ Through the enforcement of means of production and manufacturing practices, such as a minimum thermal treatment to be applied at coldest spot of the product.

As an example, a minimum sterilization value ($F^{121.1°C}_{10°C}$) of 3 minutes was widely used in most countries, and is still strictly enforced by some regulations, that is, in the 9 Code of Federal Regulations enforced by the United State Department of Agriculture, for uncured meat products (see 9 CFR, part 317).

This sterilization objective of $F_0 = 3$ minutes is based on the equivalent time of 12 log reductions of *Clostridium botulinum* spores, (assumed to have a $D_{121.1°C}$ thermal resistance of about 0.2 minutes, and a Z value of ~10°C). It does not take into account the nature and the amount of initial contamination, the ability of the surviving microorganisms to grow in the product, or the actual thermoresistance of the spores in the product.

The minimum $F_0$ level actually required for the stabilization of most food products is usually higher than 3 minutes, due to the presence of thermophilic spores with high thermal resistance, such as *Geobacillus stearothermophilus*. It can be lower than 3 minutes, in the case of nitrite salt containing products, or high fat content meat products (foie gras as an example), or for foods with an intermediate pH and a high salt content (such as olives). Thus, setting of a minimum $F_0$ value alone is not sufficient to ensure the stability of a retorted food. In these cases, a criterion such as the one shown below is used:

- Through the requirements of "stability" of the food, ensured by submitting the retorted product to stability tests.

Stability tests usually include the incubation of samples for several days at low (15–25°C), medium (30–40°C), and/or high (50–60°C) temperatures. The incubation test allows checking to see if some mesophilic and/or thermophilic microorganisms are present and capable of growing in the product.

Growth of microorganisms is usually indirectly observed through the following:

(a) The production of gases (mainly $CO_2$, and sometimes $H_2S$ or $H_2$) in the container: nonstability is declared if container deformation is observed or an increase in pressure is seen. Some microorganisms produce only a small amount of a gas during incubation and growth, and it is therefore necessary to combine the gas production test with other means of control.
(b) The decrease in pH during incubation: the pH value of the incubated sample should be no different (maximum of 0.5 pH unit difference) than the pH of nonincubated sample, kept at room temperature. However, some microorganisms produce only few organic acids during incubation and growth. In case of a decrease in pH (e.g., in the range 0.3–0.5 pH units), microscopic

examination on the sample is used as the third means of checking the stability of the product.

As an example, the French regulation related to canned food, enforces a stability test as follows (AFNOR Methods):

*Sampling of 3 or 5 containers from the production batch of retorted products:*

- control sample kept at 20°C;
- one or two samples incubated at 37°C for 7 days, or at 32°C for 21 days, then cooled at 20°C;
- one or two other samples incubated at 55°C for 7 days, and then cooled at 20°C.
- *Containers examination*: Containers should display no apparent deformation.
- *pH measurement*: No difference, equal or higher than 0.5 pH unit, between the control sample and 37°C incubated sample, is required. Examination of the food is also performed (normal color and normal aroma are required).

If all characteristics of the samples incubated at 37°C are similar to that of the control sample, stability is declared and the food safety is assumed to be ensured for the complete batch of retorted product. Nonstability observed at 55°C should be considered by the manufacturer as the result of lack of hygiene during fabrication, but it has no relevant impact on food safety. Corrective measures must be applied by the manufacturer in order to keep the products stable after 55°C incubation.

It can be seen that:

(a) Stability test is performed on a small sample of retorted products, sampled from a large population of products. The stability test has nonstatistical significance, due to the very small ratio between the sampling population and the total population. For practical reasons, only five products are sampled, preferably in the coldest zone of the retort.

(b) Stability of the food during incubation test does not mean absolute sterility of the products. For example, the stability test is performed the same way for acid or acidified foods (pH lower than 4.5), that usually withstand only a pasteurization thermal treatment—the surviving spores are inhibited from germination and growing during incubation by low pH.

The good practices for the design and validation of thermal processes detailed below are based on the translation of the French canners guidelines. They apply to different situations encountered in the canning industry:

- Prepackaged retorted foods or continuously treated bulk products, associated with aseptic packaging.
- Acid or acidified foods, or low-acid foods.

The guideline describes critical points to take into account when designing or validating thermal processes. It is the result of a technical HACCP procedure of the retorting process, leading to development of the details for means of control and methodologies.

The Guideline applies to retorted foods as defined by the French regulation. The guideline was written to apply to all packaged products, associated with thermal treatment suitable to allow room temperature stability, including products displaying only partial tightness to gases (e.g., retorted foods packaged in plastic packaging, pouches, trays, etc.).

## 6.3. Critical factors related to the design of thermal treatments, for the products packaged prior to treatment

The critical factors listed below are of importance in the determination of a scheduled process, and should thus be controlled for proper application.

### *6.3.1. Factors related to the product (as evaluated just before thermal treatment)*

#### *6.3.1.1. The microbial load*

Depending on its nature, origin, conditions of receiving, transportation, storage, cleaning, preparation, and sorting, the raw material carries different flora, in various amounts. Practically, it is very difficult to estimate the microbial load of one product. However, it is important to evaluate the global initial microbial load and the contamination of the product in the whole processing line. This evaluation can be performed by trained technicians with a good knowledge of the product, through contamination measurements on the line, or afterwards through stability checks.

The following two parameters should be assessed:

1. The destruction speed of microorganisms when exposed to a given lethal temperature (thermal treatment needed for inactivation should vary drastically, not only from one species to another, but also between different strains of same species).
2. The level of microbiological contamination at the beginning of thermal treatment (usual and satisfactory thermal treatment can become insufficient, due to excessive increase of initial microbial load).

*6.3.1.2. Determination of the thermal treatment intensity target*

The thermal treatment should ensure the biological stability of the product at room temperature, through the destruction or inhibition of all vegetative or spore-forming microorganisms, and especially *C. botulinum*, which is the target organism among pathogens. A thermal treatment is defined by time–temperature association (sterilization or pasteurization scheduled process). The scheduled process is designed for one given product, packaged in one given container, and for specific equipment characteristics.

### *6.3.2. Sterilization or pasteurization value*

The intensity of thermal treatment can be expressed as sterilization value (SV or $F_0$) or pasteurization value (PV) achieved at the critical point of the product, or for the fraction of the product undergoing the least treatment. It is expressed as time-equivalent (minutes) at a reference temperature (121.1°C for sterilization and 70°C or 93.3°C for pasteurization).

For some products, monitoring of critical point is difficult to perform. In this case, a higher SV can be chosen as the target, through a temperature monitoring performed in a less unfavorable but more easily accessible location. Monitoring of this new SV should ensure sufficient SV at the critical point, and therefore ensure biological stability of the product at room temperature. For example, a $10^{12}$ reduction of *C. botulinum* spores count is obtained for a minimal sterilization value of $F_o = 3$ minutes.

Practically, SV or PV is chosen to ensure the destruction or inhibition of the nonpathogenic but more heat-resistant microorganisms, likely to spoil the product, such as some strains of *Bacillus* coming from natural flora of vegetable products, and also considering the history of the firm. No minimal SV or PV is required from European and French regulations, and thus, biological stability of the retorted food is the only regulatory requirement.

#### *6.3.2.1. Product characteristics that determine the design of thermal treatment*

##### 6.3.2.1.1. The pH of the product

Depending on the pH of product, two different stabilization treatments are distinguished:

1. Naturally acid, or acidified foods, with a pH equal to or lower than 4.5, in which most bacteria cannot proliferate, are subjected to thermal treatments at temperatures usually below 100°C, commonly called pasteurization.
2. Low-acid foods (pH higher than 4.5), allowing the development of heat-resistant bacteria, are subjected to treatments above 100°C, commonly called sterilization.

##### 6.3.2.1.2. The water activity ($a_w$) of the product

The water activity is a measurement of the fraction of water available in a product, for the growth of microorganisms. Decreasing $a_w$ inhibits microbial growth. Low-acid and high moisture content products, with $a_w$ higher than 0.85, should be subjected to thermal treatment, in order to become stable at room temperature.

##### 6.3.2.1.3. The nature of the product (convective or conductive heating behavior)

The nature of the product (viscosity, particle size, and solid/liquid ratio) is a very important factor since heat is transmitted to the product (during the thermal stabilization process) through the container wall, to the product, up to the slowest heating zone, also called the critical point. The critical point or zone should be subjected to the minimal required thermal treatment. Thus, all remaining zones of the product undergo a thermal treatment, at least equivalent to this value. The temperature at the critical point, versus time, should be selected as the reference for the monitoring and evaluation of the scheduled process. In the case of nonhomogenous products, the critical point is usually located in the largest particle or the lowest heat-conductive particle, which is not necessarily located at the geometrical center of the package.

Additives and ingredients used in product formulation and for product pretreatments, are likely to influence product behavior during thermal treatment (i.e., increase in viscosity) and the heat resistance of microorganisms. The nature and exact amount of additives and ingredients should therefore be taken into account.

### *6.3.3. Influence of pretreatments (blanching, cooking)*

During the preparation of the product, all pretreatments such as blanching or cooking prior to final thermal treatment, should be adopted and controlled, according to the cooking effects in the forthcoming pasteurization or sterilization step. Preheating or precooking of foods allow evacuation of gases present in the products and containers, which expand during thermal treatment, and produce excessive pressure, which is detrimental to container tightness. Degassing is required for products containing a lot of entrapped gases such as fruits, vegetables, or meat products.

### *6.3.4. Factors related to the packaging*

Based on their nature and on the closure technologies used, packaging used for shelf-stable products should comply with technical requirements: mechanical resistance and barrier properties, especially toward microorganisms.

#### *6.3.4.1. Nature and size of packaging*

Heat penetration in the product depends both on the nature of the product itself and on the packaging—shape, nature of material, size, and filling level. Changing the packaging necessitates changes in the thermal treatment to be used.

*6.3.4.2. Influence of the filling level of packaging*

Filling the containers should be controlled, since excessive filling is likely to modify the thermal treatment unfavorably, to be detrimental to tightness (through the deformation of the packaging), and to lower the convective currents, thus reducing the rate of heat transfer. It is also of importance in particulate foods where the liquid to solid heat transfer may be minimized. For all processes with rotation of the containers, control of the head-space volume is essential, since it determines the size of the void "bubble," the displacement of which may enhance heat transfer in the container.

### *6.3.5. Factors related to the process*

The basic principle of sterilization uses three parameters—temperature, pressure, and time. Precise application of these parameters must be ensured by suitable thermal treatment equipments (batch or continuous retorts, still or agitating, and by the nature of the heating medium—steam or hot water).

*6.3.5.1. Rotation or agitation*

Rotation or agitation of the containers (axially or "end over end," or other motions) during thermal treatment, enhance, for some products, the heat penetration into the product, by inducing forced convection. The efficiency of rotation is determined by several factors such as rotation speed, head-space volume, solid/liquid ratio, particle size, and product consistency/viscosity. If agitation has been scheduled when designing the process, it must take into account both rotation speed and position of the containers (for either axial or "end over end" rotation).

*6.3.5.2. Basket loading pattern*

Proper circulation of the heating medium is largely responsible for good homogeneity of temperature, as required in the retort. It is very important that the heating medium (water, steam, or steam + air) may circulate homogeneously between all containers. Thus, a loading pattern for the baskets in the retort must be defined when designing the process—use of separators, stacking rules, and so on.

*6.3.5.3. Initial temperature*

The initial temperature of the product should be controlled for proper sterilization. An initial temperature lower than the scheduled value may have a detrimental effect on the intensity of thermal treatment in the product. Initial temperature may vary, depending on the control of the initial temperatures for each food ingredient when filling the container, and may also be related to the standby duration between the time of loading of the first container and the last container in the retort (for batch processing).

*6.3.5.4. Come-up time (CUT)*

The come-up time (time for the retort to reach retorting temperature) should be noted when designing the scheduled process. If the actual come-up time is shorter than expected, the rise in temperature in the product may be lower during this step. The nominal come-up time of the production equipment should be used and it should be determined under production conditions (usual load of the retort). A scheduled process designed for a given CUT should not be used if the actual duration of CUT is shorter. It should be noted that some continuous processes have a zero CUT.

*6.3.5.5. Retorting temperature*

The scheduled retort temperature during thermal processing should be controlled for each step of the process. A temperature lower than the scheduled set point, may reduce the efficiency of thermal treatment and pose a hazard to food safety.

*6.3.5.6. Cooling*

The temperature profile during the cooling phase (temperature of cooling fluid and flow rate) should be controlled in order to ensure proper application of the thermal treatment, since a portion of the SV or PV is obtained, for some products, during the cooling phase. Thus, excessively cold cooling water may enhance and shorten the cooling of the product, thereby reducing the global intensity of the thermal treatment. On the contrary, a high temperature for the cooling water may induce degradation of the product final quality. Cooling conditions should thus be carefully monitored and determined when designing the process. Water treatment may be performed in order to ensure the sanitary quality of the cooling water.

*Note*: The control of counter-pressures and heat transfers during thermal treatments, including the cooling phase, is essential to ensure the integrity of packaging.

### *6.3.6. Skill of employees*

The employees in charge of designing and validating the thermal treatments should have the skill and ability to master all parameters, related to the design and adjustment of sterilization or pasteurization processes, including product-, packaging-, or equipment-related parameters.

## 6.4. Critical factors related to the design of thermal treatments, for products treated prior to aseptic packaging

### *6.4.1. Specific factors related to the product (as evaluated at the inlet of heating section)*

The nature of the product (viscosity, particle size, and solid/liquid ratio, heat, and/or electrical conductivity) is a very important factor, since heat transfer or heat generation in the equipment directly depends on the characteristics of the product and on the resulting flow pattern (laminar or turbulent flow) at a given flow rate. Bulk heat treated products should withstand pumping throughout the stabilization equipment, up to the aseptic packaging device. Blanching or precooking of foods is often required prior to heat treatment of pumpable products, allowing evacuation of all gases likely to be detrimental during bulk heat treatments, since the presence of gas bubbles leads to unstable flow in continuous devices.

### *6.4.2. Specific factors related to the process*

The different available technologies for bulk heat processing can be divided into:

- *Indirect heating*: Tubular, scraped surface, plate, or other heat exchangers allow the product to be progressively heated while keeping it in close contact with a heated surface.

- *Direct heating*: Ohmic heating and microwave heating use some characteristics of the product (electrical conductivity, water content, and dielectric properties) to ensure fast and direct heating through heat generation in the product. Other heating methods use food-grade steam injection directly in the product.

*6.4.2.1. Initial temperature*

Depending on the equipment configuration, the initial temperature of the product, at the inlet of heat treatment section, is either the temperature at feeding pumps, or the temperature adjusted after flow through preheating section. Initial temperature is very important if the intensity of heat treatment during the flow is fixed, or regulated only based on the outlet product temperature.

*6.4.2.2. Residence time distribution (RTD)*

For all heat processes applied to bulk food products, measuring the temperatures actually encountered by the product through all equipment is not possible. The intensity of actual treatment is not homogeneous and can only be estimated on the different product fractions by the combination of:

- the residence time distribution of the product particles and continuous phase, in the heating, holding, and cooling zones;
- the temperature profile in the equipment.

RTD is influenced by the device geometry, product flow rate, and rheological characteristics of the product. In addition, for processes that use the Joule Effect (ohmic heating), variation in electrical conductivity of product components is likely to directly result in wide RTD and hence a nonhomogeneous treatment.

*6.4.2.3. Cooling profile*

The temperature profile in the cooling section (temperature of the cooling fluid and flow rate) should be controlled in order to ensure proper application of the thermal treatment, since a portion of the SV or PV is obtained, for some products, during the cooling phase. The RTD in the cooling section should be taken into consideration when designing the process.

## 6.5. Qualification of heat stabilization equipments

### *6.5.1. Retorting and pasteurization equipments for packaged food products*

#### *6.5.1.1. Batch retorts or pasteurizers*

The heat treatment of sterilization batch equipments should allow to ensure, for each individual container, proper application of a scheduled process, whatever its location in the equipment. Heat distribution studies should therefore be performed with appropriate repetitions, in order to locate the coldest zones of the retort. A scheduled process is thus designed to ensure proper SV or PV at the critical point for products located in this zone and subjected to the least heat treatment.

Advised procedure:

- Simultaneous temperature measurement and recording in different locations of the retort's baskets (at several levels from top to bottom) should be done, using temperature probes or wireless dataloggers. The number of measurement points and their locations must be adjusted according to the size of the retort.
- Heat distribution studies should be performed with a full load of the equipment, under normal production conditions, and using the predetermined packaging.

The recorded time–temperature data can be analyzed in different ways:

- *Temperature range during the "constant temperature" phase (usually called the "scheduled process")*: The maximal and minimal temperatures are recorded during the process, and their respective locations are noted. The location of the lowest temperature observed is assumed to be the critical zone of the retort. This often leads to inappropriate determination of critical zone, since industrial equipments are never as precise as expected, and also because it does not take into account the evolution of temperature with time—the minimal temperature at one moment does not reflect the minimum intensity of the whole process at this location. Moreover, the duration of the scheduled process is sometimes not precise.

- *Average temperature during the "constant temperature" phase*: This calculation leads to locating the "lowest temperature zone" during the heat process, but does not reflect the actual intensity of the whole process. For processes with a long "come-up time," a part of the sterilization treatment is delivered before the constant temperature heating phase occurs. Also, for nonconstant temperature processes, this method is not relevant.

A new approach, in order to correctly locate the critical area of a batch equipment, was developed in addition to the requirements described in the "Guideline for the heat distribution study in batch retorts" designed and edited by the FDA (this guideline was initially described for water cascading retorts, but can be used for other heating media and retort technologies):

We introduced the use of Global $F^{121.1°C}{}_{10°C}$ achieved in the retort ambient medium during complete sterilization cycle (CUT, heating, and cooling phase). This additional indicator, designated as "Sterilization Power Integrated Intensity," or SPII, reflects the ability of the heating medium to transfer heat to the packaging, and also takes into account eventual lack of homogeneity during the cooling phase, due to distribution of the cooling fluid (which is not always similar to the heating medium distribution).

The integrative effect enhances differences, or points out compensations between the heating and cooling phases, so that the total sterilization process homogeneity can be evaluated. The location of the "critical zone" of the retort is therefore not always the location of the "slowest heating zone," but more precisely takes into account the location where the total $F_0$ in ambient medium, SPII, is minimal. This method of use of recorded temperature data was accepted by FDA for the heat distribution study of water cascading retorts, in which sometimes strong differences in SPII were observed between the top and bottom of the baskets.

Table 6.1 illustrates an example of the use of SPII in a heat distribution study for a thermal process.

The overall range of average temperatures observed during the heating phase (excluding CUT), is 1.0°C (highest: 124.8°C, level 1 of basket 2; lowest: 123.8°C, level 3 of basket 4). From these observations, the "coldest zone" of the retort should be located at

**Table 6.1.** Heat distribution study in four baskets in a retort (performed with a full load of packages)

| Location of datalogger in the retort | Average temperature during process (°C) | Partial $F_o$ in retort during CUT (minutes) | Partial $F_o$ in retort during scheduled process (minutes) | Partial $F_o$ in retort during cooling (minutes) | SPII: global $F_o$ in the retort during the process (minutes) |
|---|---|---|---|---|---|
| *Basket 1* | | | | | |
| Level 1 | 124.6 | 0.3 | 12.7 | 2.4 | 15.4 |
| Level 2 | 124.3 | 0.3 | 12.4 | 2.6 | 15.3 |
| Level 3 | 124.3 | 0.4 | 12.1 | 3.1 | 15.6 |
| *Basket 2* | | | | | |
| Level 1 | 124.8 | 0.4 | 13.0 | 2.7 | 16.1 |
| Level 2 | 124.6 | 0.3 | 12.7 | 2.5 | 15.5 |
| Level 3 | 124.5 | 0.3 | 12.5 | 2.6 | 15.4 |
| *Basket 3* | | | | | |
| Level 1 | 124.4 | 0.2 | 12.8 | 2.2 | 15.2 |
| Level 2 | 123.9 | 0.3 | 12.6 | 2.3 | 15.2 |
| Level 3 | 124.0 | 0.2 | 12.3 | 2.3 | **14.8** |
| *Basket 4* | | | | | |
| Level 1 | 123.9 | 0.3 | 12.3 | 2.3 | 14.9 |
| Level 2 | 124.0 | 0.4 | 12.2 | 2.6 | 15.2 |
| Level 3 | **123.8** | 0.4 | 12.1 | 2.5 | 15.0 |

level 3 of basket 4. However, the lowest recorded SPII shows that level 3 of basket 3 is the actual "critical zone" of the retort. Thus, heat penetration trials, and sampling for stability assessments, should be performed in this area.

*6.5.1.2. Continuous sterilizers*

Continuous sterilization equipment (rotary continuous sterilizers, hydrostatic sterilizers, pasteurization tunnels, etc.) should allow to ensure, for each individual container, proper application of the minimal scheduled process, whatever its actual residence time in the equipment. Heat distribution and/or RTD studies should therefore be performed with appropriate repetitions, in order to estimate the conditions of the process for the fraction of products withstanding the lowest intensity of heat treatment.

Advised procedure:

- Conduct simultaneous temperature measurements and recordings at different locations of the packaging conveying device, using wireless dataloggers. The number of measurement points and their location must be adjusted according to the geometry of the equipment. Measurements should be repeated several times, to evaluate the stability of temperatures during processing.
- Heat distribution studies should be performed with a full load of the equipment, under normal production conditions.

The use of recorded temperatures can be handled as described for batch retorts, followed by determination of the SPII.

### 6.5.2. Retorting and pasteurization equipments for bulk-treated foods, associated with aseptic packaging

*6.5.2.1. Validation of presterilization and absence of biological risk through recontamination*

The sterilization cycle of the processing line, prior to production, should be designed to ensure suitable decontamination in each location. Sterilization frequency must be adapted according to the production running time, and to fouling due to product flow.

*6.5.2.2. Qualification of regular product flow delivered by the feeding pump*
Due to the direct relationship between flow rate and RTD, the regularity of feeding should be ensured periodically.

*6.5.2.3. Fouling of the equipment*
Designing and scaling of the equipment must be adapted to the product nature, and thus fouling (by peripheral overheating of product) can be limited, therefore ensuring an appropriate and similar RTD.

*6.5.2.4. Determination of residence time distribution*
Calculation of flow rate and velocity profile in the geometry of the processing unit (and scaling-up from pilot-sized equipment), enables us to determine the theoretical RTD in it. The worst case time (fastest fluid element) is assumed to be half the average time corresponding to the flow rate (perfect laminar flow). However, it is strongly recommended to perform experimental measurements, using salt tracers and conductivity measurements, for example, in order to evaluate actual RTD of the fluid and particles, especially the minimum residence time, which is used for the calculation of the required process.

## 6.6. Design and validation of thermal treatments

The validation of one newly designed thermal treatment, for food products either packaged before of after heat treatment, should be performed through precise methodology, such as by the methodology described in Figures 6.1 and 6.2, or by any other method giving similar results for ensuring food safety.

When designing a new product and its thermal process, all critical factors of any relevance in the production operation should be taken into consideration to ensure food safety. However, scaling-up from laboratory or pilot-scale to an industrial scale may have consequences on critical factors, with difficulty of control. Thus, performing pre-trials permit checking the validity of the scheduled processes, before running the actual production run. All data collected during different steps of the validation process should be recorded in one validation file, and kept in the archives of the manufacturer.

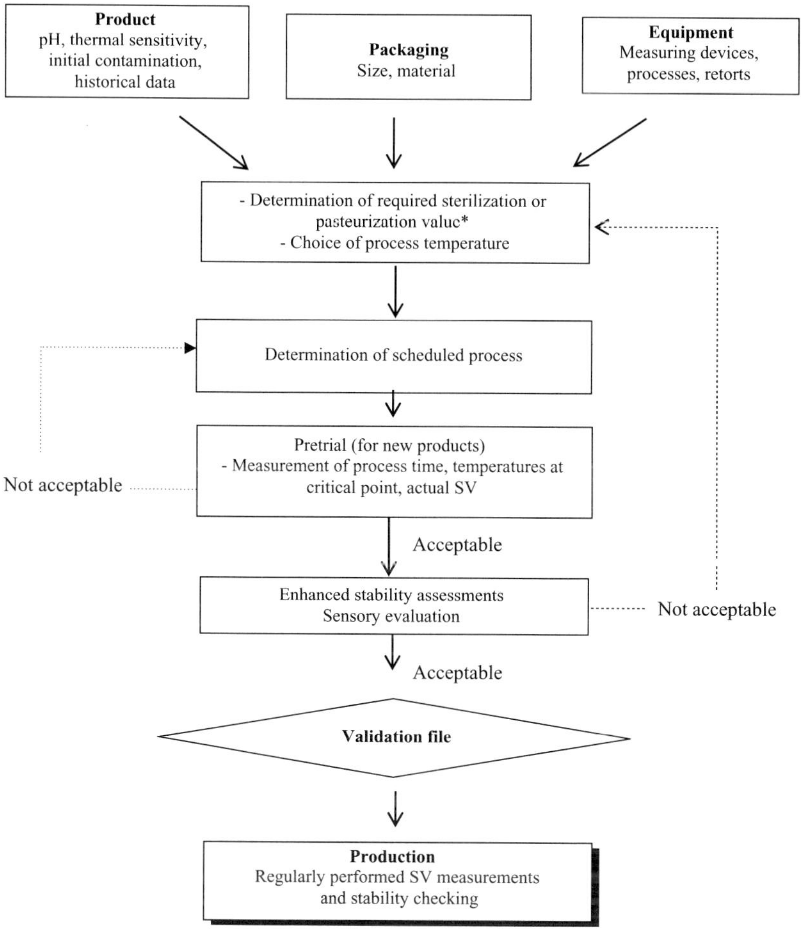


*Required sterilization value = minimum value to be achieved at critical point, to ensure stability at room temperature

**Figure 6.1.** General methodology for the validation of sterilization or pasteurization scheduled process, for products packaged prior to thermal treatment.

## 6.7. Heat destruction parameters and sterilization value

It is assumed that actual microbial spores display a thermal resistance behavior similar to that of a first-order reaction (Ball and Olson, 1957; Stumbo, 1973) with the influence of temperature governed by classical Arrhenius' laws:

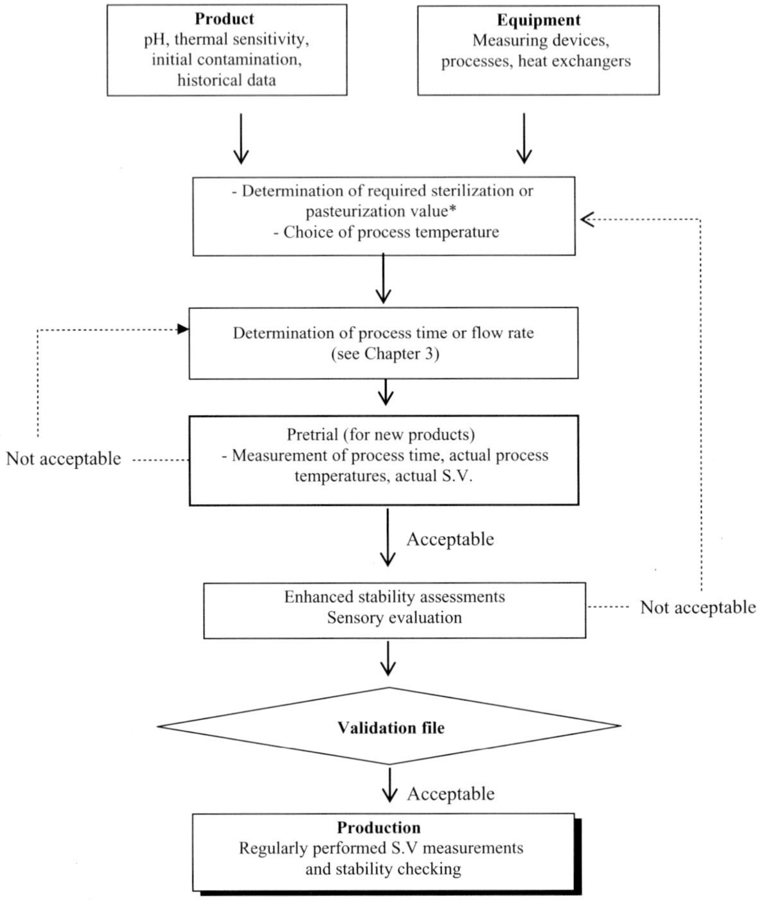


*Required sterilization value = minimum value to be achieved for the fraction of product experiencing the shortest residence time, to ensure stability at room temperature.

**Figure 6.2.** General methodology for the validation of sterilization or pasteurization scheduled process, for products subjected to bulk thermal treatment prior to aseptic packaging.

The influence of the heat treatment time and temperature can be expressed as:

$$t = D_T \log \frac{N_0}{N} \tag{6.1}$$

$$\log \frac{D_{T_{\mathrm{ref}}}}{D_T} = \frac{T - T_{\mathrm{ref}}}{Z} \tag{6.2}$$

with:

$t$ Total duration of the heat treatment at constant temperature $T$

$D_T$ Duration of decimal reduction (time for 90% destruction of the target, at the temperature $T$)

$N_0$ Initial number or amount or concentration of the target (in practice: initial number of microbial spores)

$N$ Final number or amount of the target after heat treatment (in practice: final count or "probability of presence" of surviving microbial spores)

$T_{ref}$ Reference temperature chosen as "central point" for comparison of treatments (in practice: for sterilization treatments—121.1°C (250°F); for pasteurization treatments—70°C or 93.3°C; for evaluation of cooking effects—100°C)

$D_{T_{\text{ref}}}$ Decimal reduction time (usually expressed in minutes) at the reference temperature (as an example: $D_{T_{\text{ref}}}$ of *G. stearothermophilus* in most vegetables products is in the range of 3–6 minutes)

$Z$ Temperature Activation Factor (expressed in °C): increase in temperature that lowers the D value by 90%

The $Z$ factor is directly related to the Activation Energy used in Arrhenius' laws (e.g., $Z$ is usually in the range 5–15°C for the destruction of vegetative microbial cells and of microbial spores; Z is in the range 20–40°C for the effects of heat on biochemical characteristics of food: destruction of vitamins, texture losses, etc.)

The two parameters $D_{T_{\text{ref}}}$ and $Z$ are necessary and sufficient to characterize completely the heat resistance of one microbial strain in a given medium.

A graphical representation of the heat destruction behavior is widely used by canners and is displayed in Figure 6.3.

To ensure easy comparison of different treatments, both isothermal or nonisothermal, the SV (or $F_0$) is defined as a "heat treatment intensity scale":

(a) It can be assumed that microbial spores are characterized by a $Z$ value close to 10°C.

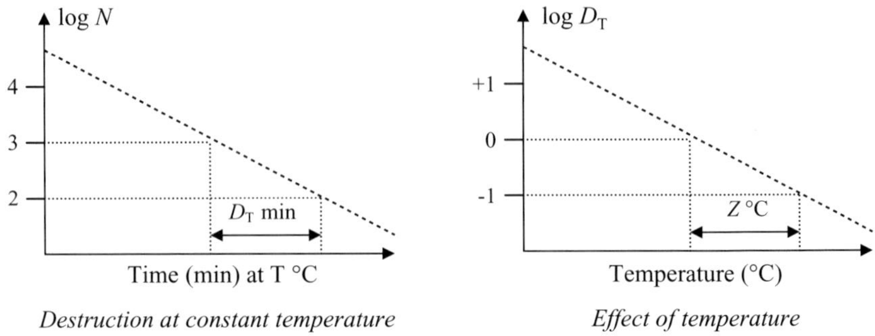


**Figure 6.3.** Basic kinetic parameters used for sterilization processes: first-order heat destruction behavior and graphic representations.

(b) A reference temperature of 121.1°C is chosen based on international agreement.
(c) The total integrated effect of one nonisothermal treatment can be expressed as the additive effect of each time at different temperatures.

Thus, it is possible to express the SV as:

$$F_o = D_{121.1} \log \frac{N_0}{N} \tag{6.3}$$

and

$$F = t \times 10^{\frac{T - T_{ref}}{Z}} \text{ for a constant temperature process} \tag{6.4}$$

$$F = \int_0^t 10^{\frac{T - T_{ref}}{Z}} dt = \sum 10^{\frac{T - T_{ref}}{Z}} \Delta t \text{ for a variable temperature process} \tag{6.5}$$

Equation (6.5) with classical parameters is known as the Bigelow equation, and is widely used for the calculation of total $F_o$ achieved at the cold point during industrial sterilization treatments.

### *6.7.1. Calculation of process duration using the Ball method*

In order to understand how the semi-empirical Ball method is used to calculate the duration of heat process, according to the various requirements, it is useful to present a classical batch retort cycle one product heated by conduction, as displayed in Figure 6.4.

In Figure 6.4,

| | |
|---|---|
| $T_r$ | Retort temperature (°C) |
| CUT | Come-up time; time required to raise the retort temperature to the set point (min) |
| $T_o$ | Initial temperature of the product (°C) |
| $B$ | Scheduled process duration (min) |
| $T_g$ | Temperature achieved at cold point at the end of heating phase (°C) |
| $F_o$ | Total sterilization value achieved at the end of cycle (heating then cooling phases): $F_o = F_{o\text{-h}} + F_{o\text{-c}}$ |
| $F_{o\text{-h}}$ | Sterilization value achieved during heating |
| $F_{o\text{-c}}$ | Sterilization value achieved during cooling |

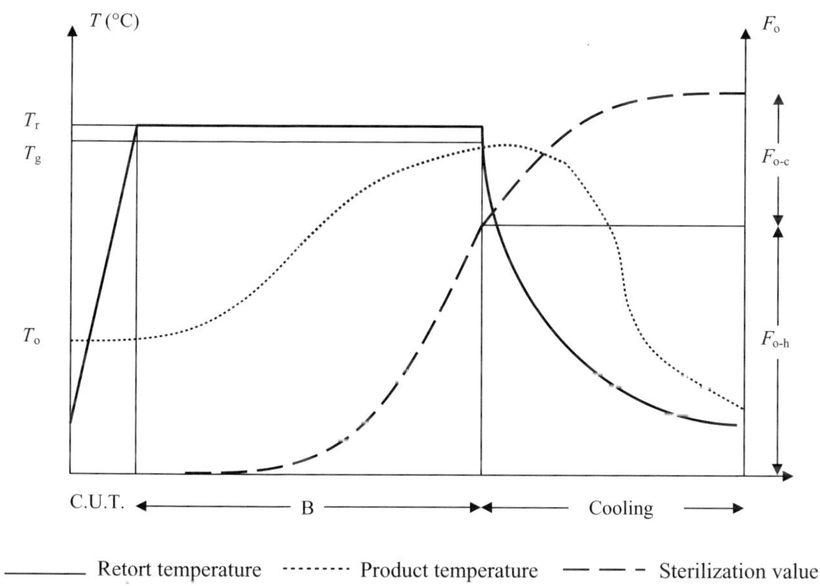


**Figure 6.4.** Evolution of temperatures and sterilization value during retorting cycle.

The Bigelow method (integration of product temperature during the process) can only be used to calculate the $F_o$ under a given retorting condition, but cannot be used for the calculation of required duration ($B$), to achieve a desired $F_o$.

For the calculation of the scheduled process duration ($B$), it appears necessary, from this diagram, to supply several input parameters such as: $T_r$, $T_o$, required total $F_o$, equipment characteristics (CUT), cooling water temperature, and the heat penetration parameters of the product and package under the conditions of retorting (still or agitating conditions).

The heat penetration parameters most widely used in canning are those defined by Ball and Olson:

- The $f$ parameters ($f_h$ during heating and $f_c$ during cooling, expressed in minutes) are characteristics of the heat transfer "speed," from the retort to cold point of the packaged product. They are related to both thermal diffusivity and/or the rate of convective heat transfer, and of the shape & size of the packaging.
- The $j$ parameter (without dimension) is related to the convective or conductive heating behavior of the product.

Using those two parameters, which have to be obtained through experimental measurements, and all other previously noted process parameters, the Ball method leads to an evaluation of the required process duration ($B$).

A relationship was established by Ball between $F_o$, $T_o$, $T_r$, $f_h$, $j$, and the resulting parameters $T_g$ and $B$. Very simple formulae and empirical tables lead to proper evaluation of $B$.

The main advantages of the Ball method are as follows:

- In most cases, it leads to thermal processes that are very suitable for packaged foods with a convective heating behavior.
- The scheduled process ($B$) can be easily corrected for each value of CUT.
- It is possible to calculate $f_h$ and $j$ values obtained from one experimental run in one product geometry, for slightly different geometries (e.g., from 1/1 cylindrical can to 1/2 cylindrical can). It is therefore easy to calculate the required process time for other can sizes.

- Complete "families" of scheduled processes can be quickly calculated for various product parameters and retort conditions ($T_r$, $T_o$, required $F_o$).

However, the Ball method also presents several drawbacks and limitations:

- It is a laborious technique, due to the absolute requirement for precisely measured heat penetration parameters (preferably on several independent retort runs).
- The proper determination of $f_h$ and $j$ values absolutely requires an experimental measurement with a CUT as short as possible (less than 3 minutes in practical conditions). This cannot be achieved in an industrial retort: a special pilot-sized retort is needed.
- The Ball method only calculates one scheduled process: it remains necessary to measure and check the actual $F_o$ during the industrial process, when the calculated process is applied.
- The Ball formulas are likely to underestimate the $F_{o\text{-}c}$ obtained during cooling phase, especially in the case of strong inertial heating behavior (known as "overshooting" phenomenon) of conduction heated products: temperature at cold point usually increases over $T_g$, leading to noticeable and unnecessary oversterilization (not a drawback itself), but also severe overcooking of the product, with poor retention of nutritional and organoleptic properties. It also leads to a longer process that reduces the throughput and increases the energy consumption.

In practical applications, $F_{o\text{-}c}$ can be as high as 30%, and sometimes up to 50% of total $F_o$. Under such conditions, it is very difficult to optimize the process. Proper integration of the cooling phase, as an important part of the process, is then a challenge that requires alternate calculation methods.

## 6.8. Real-time optimization of retort process: A new approach

In order to perform optimization of thermal treatments, even in the case of strictly conductive products in large packages, a specially designed software (OPTIBAR®) was developed, to minimize the total thermal treatment of batch retorted foods, with respect to the total $F_0$ of the process.

Development of the calculation concept and software codes were conducted by CTCPA and the Thermal Engineering Research Unit, at ENSIA (Ecole Nationale Supérieure des Industries Alimentaires) by A. Duquenoy, with the help of Comeureg Company, who designed the computer/temperature probes interface, and the final user-friendly software.

The calculation of "just as necessary" duration is performed through an iterative calculation that directly uses data obtained from experimental runs during the sterilization of the product under actual industrial conditions. The optimization procedure was developed by modeling of heat transfer during thermal treatment (Duquenoy, 1983, 1991; Beyer *et al.*, 1988).

The method uses the prediction, starting from recorded data, of the forthcoming heating phase, and then the forthcoming cooling profile of the product. In order to achieve proper optimization, actual behavior of the product has to be investigated through experimentation (Figure 6.5).

The temperature at the cold point of three product samples are recorded, the cumulative $F_0$ is calculated for each sample, and results displayed in real time on the screen, through an interface and specially designed software with user-friendly windows. When the operator starts the process, the product starts to heat, and it can be easily seen that the achieved $F_0$ increases till it reaches the set point. Then it is time to stop with heating and start cooling the retort.

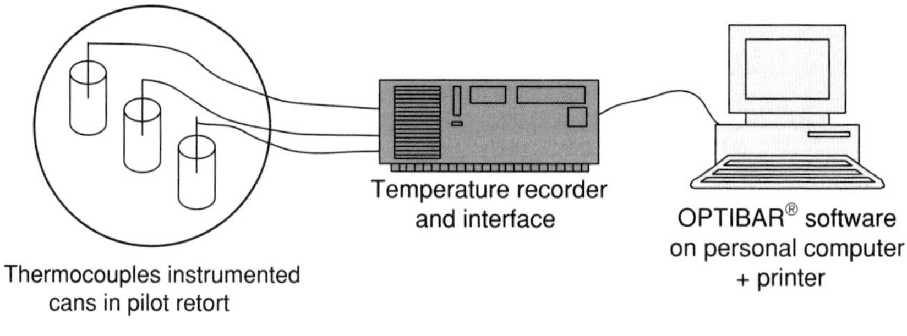


**Figure 6.5.** Experimental device for the real-time optimization of retort processes.

A step-by-step description of the optimization procedure is given below.

### 6.8.1. Phase 1: Initial data capture

The main screen window displays a set of initial data to be entered manually by the operator:

- Data for registration (name and details of the product).
- Reference temperature and $Z$ value for the calculation of integrated $F$ values. (Note that both pasteurization or sterilization processes can be optimized based on the target $F_0$ and the cooking value.)
- Target $F_0$ value.
- Conditions for the retorting: retort temperature set point, indication of CUT; cooling water temperature.
- Type and geometry of the packaging (cylindrical or square shaped).
- Product temperature and retort temperature. (Note that heat penetration parameters are not required for the computer to optimize the process.)

### 6.8.2. Phase 2: Start of the process and temperature rise

When the heating medium is introduced into the retort, the optimization procedure is launched by operator. During the beginning of the cycle, the software only records the temperature and calculates the $F_0$ for each can. The "slowest heating can" displaying the lowest $F_0$ is automatically selected and followed as the "worst case." The computer may pick another can to be the "worst case" at a later point of time, based on the heat treatment received by each can. The end of CUT is automatically detected when the retort temperatures is no more than 1°C below the set point temperature.

### 6.8.3. Phase 3: Sterilization at constant temperature

#### *6.8.3.1. Phase 3.1: Unknown constant temperature*

The first recordings are used during the stabilization of the retort, in order to evaluate the actual retort temperature, suitable for the calculation loop to start optimization. When the retort temperature is observed as stable, an average $T_r$ is calculated and displayed, and is then used for further optimization.

*6.8.3.2. Phase 3.2: Sterilization at known constant temperature*
This phase uses the "heart" of the software, including the iterative calculation loop.

- At time $t_i$, all experimental, actually recorded data of the heating rate at the cold point is used by the software:
  - to calculate $F_{o\text{-}h}$,
  - to extrapolate the forthcoming temperatures and $F_{o\text{-}h}$ to be added in the next few minutes.
- The target for $F_{o\text{-}h}$ to be achieved at end of cooling is fixed, at the beginning of optimization, to be 50% of total $F_o$.
- Time for the end of heating (50% of $F_o$ obtained) is then extrapolated.
- From this data, possible forthcoming $T_g$ is calculated, and then the cooling profile is extrapolated at the cold point. $F_{o\text{-}c}$ is the calculated, then added to $F_{o\text{-}h}$.
- The resulting $F_o$ (actually obtained $F_{o\text{-}h}$ at $t_i$ + forthcoming $F_{o\text{-}h}$ up to end of heating + forthcoming $F_{o\text{-}c}$) is compared with the target $F_o$.
- The "time for end of heating" is displaced forward or backward in time, to match the total required $F_o$.
- The calculation starts again using the last recorded data, leading to better evaluation of process time and $T_g$, until optimization is achieved in real time.

The simulation of the heating and cooling profiles are obtained by integration of the Fourier's laws or using a convection model, for usual geometries. All the needed physical parameters (including thermal diffusivity) are obtained in real time from the measurement of the product/packaging characteristics during the experiment, and thus no additional data have to be measured by the operator.

Note the following:

- As time goes by, more and more experimental data are available for the proper evaluation of the product physical characteristics. The optimization loop always uses the last collected points: Actual thermal conduction parameters are used, even when they are slightly changing during experiment. The evaluation becomes more and more precise as the temperature at the cold point becomes closer to the evaluated $T_g$ at end of heating.

- The optimization loop does not use the Ball method, but a simulation of real heat transfer in the product.
- The "overshooting phenomenon" is thus completely integrated in the model, and no overestimation of required heat treatment occurs, even for strongly inertial products.

### 6.8.4. Phase 4: "Time for cooling" and end of process

Two minutes before the heating ends, the computer gives a signal to ensure that the operator is ready for starting the cooling phase. When $T_i$ at cold point reaches the evaluated required $T_g$, signal is given for immediate cooling.

The collection of experimental data ($T$, $F_0$) runs until the end of cooling. A short file including condition of the trials and the final results (process duration and process temperature, actual $F_0$ achieved for each sample and the target $F_0$) and graphs are also provided.

The combination of experimental measurement of the actual thermal behavior of the product and the simulation in real time of the forthcoming process gives good results in predicting the required duration of the heating phase when all parameters are correctly mastered (especially a uniform initial temperature of the product).

It has to be noted that the optimization procedure gives only one scheduled process, usable only under the industrial conditions that have been used for its determination:

- retort temperature, heating medium, CUT, and cooling profile;
- product formula, packaging size and geometry, filling level in packaging;
- initial product temperature.

No other processes can be derived from the optimized one and no corrections can be made afterwards

For this reason, the OPTIBAR® software was designed for the development of processes before running into production. So it is strongly suggested to perform the optimization using the "worst cases" for all process parameters. The evaluation of the performance of the software was conducted on a complete range of industrially canned products, using water immersion as the heating medium, as displayed in Table 6.2.

Note that SV or PV, and retorts temperatures in Table 6.2 are examples developed for research only and should not be used for industrial production.

Using the optimization procedure, the actual $F_0$ obtained are very close to the expected ones. Errors do not exceed 5–10% even for products known as "difficult to optimize": low $F_0$, strong inertial behavior with overshooting phenomenon, high retort temperature, short process, and so on. This accuracy is better than the results obtained using the Ball method. No actual $F_0$ values were lower than expected ones, and so food safety is completely ensured with the use of this optimization software.

In the case of ravioli in tomato sauce, precooked ravioli is packed with a large amount of very fluid sauce. Swelling of ravioli during heating makes the initial low $f_h$ increase drastically during the last few minutes. The scheduled process should be considered as "preheating," and more than 80% of the total $F_0$ is achieved during cooling phase, which is slow. Through the iterative software, the process is optimized using the relevant data collected at the end of heating phase when the heat penetration has slowed down.

The limitations of the OPTIBAR® software occur in the following cases:

- The process is developed with a combination of high retort temperature and low $F_0$ value or convective heating behavior of the product. If the required $F_0$ is achieved only a few minutes after the end of CUT, the iteration loop has no time to perform the optimization. In this case, lowering the retort temperature is the best solution.
- The retort temperature is not stable.
- The cooling profile is not stable, or not known by the operator.

This optimization technique was used in our laboratory to screen different retort temperatures for the same total $F_0$, with a parallel

**Table 6.2.** Optimization of scheduled processes using a real-time iterative software

| Product | Retort temperature (°C) | Target sterilization (or pasteurization) value (min) | $F_{o\text{-}h}$ achieved during heating (min) | Total $F_o$ (min) | Error (%) |
|---|---|---|---|---|---|
| Mixed Mediterranean vegetables (1/1 high can) | 125 | 8 | 5.2 | 8.3 | + 3.8 |
| Beef tongue in sauce (1/2 can, low size) | 115 | 7 | 5.2 | 7 | 0 |
| Sliced potatoes in cheese sauce (1/1 high can) | 115 | 4 | 1.8 | 4.1 | + 2.5 |
| Ravioli in tomato sauce (1/1 high can) | 127 | 12 | 2.2 | 12 | 0 |
| Chili con carne (1/1 high can) | 124 | 6 | 1.9 | 6.1 | + 1.7 |
| Beef in wine sauce with vegetables (1/2 high can) | 125 | 6 | 1.5 | 6.4 | + 6.7 |
| Cassoulet—baked beans with sausages (aluminium tray, 200g) | 122 | 7 | 4.6 | 7 | 0 |
| Foie gras (glass jar, 370 ml) | 108 | 3 | 2.8 | 3 | 0 |
| Foie gras (glass jar, 270 ml) | 85 | (200) | (109.5) | (204.5) | + 2.3 |
| Foie gras (1/4 medium size can) | 82 | (175) | (100.8) | (175.9) | + 0.5 |

exploitation of temperature data for the calculation of cooking values. This leads to the choice of the best temperature for retorting (time–temperature combination giving the minimum cooking value) for a given $F_o$ at the cold spot.

## References


Ball, O. and Olson, W. 1957. *Sterilization in Food Technology*. Mc Graw-Hill Book Co., New York.

Beyer, O., Duquenoy, A., Couvrat Desvergnes, B., and Carrier, A. 1988. Vers l'ajustement automatique des baremes de stérilisation. In: *Automatic Control and Optimisation of Food Process*, M. Renard and J.J. Bimbenet (eds.). Elsevier Applied Science, London, pp. 223–239.

Codex Alimentarius (1993) Recommended International Code of Hygienic Practice for Low and Acidified Low Acid Canned Foods. CAC/RCP 23-1979. Rev. 2. (http://www.codexalimentarius.net/download/standards/24/CXP_023e.pdf).

Duquenoy, A. 1983. Duquenoy, Modelization des Transferts de Chaleur lors de la Sterilisation d'une Conserve. In: *These pour docteur-ingenieur*. ENSIA, Massy, France.

Duquenoy, A. (1991). Pénétration de la chaleur et calcul des barèmes de stérilisation. In: *La Conserve Appertisée: aspects scientifiques, techniques et économiques*, J. Larousse (ed.). Tec & Doc Lavoisier, Paris, pp. 207–231.

Journal Officiel de la Republique Francaise (JORF), 1955. Réglementation de base concernant toutes les conserves et semi-conserves alimentaires en récipients hermétiques (lait et produits laitiers exceptés). Décret n°55-241 du 10 Février 1955, J.O.du 13/02/55.

Stumbo, C.R. 1973. *Thermobacteriology in Food Processing*, 2nd ed. Academic Press, New York.

# Chapter 7

# INSTRUMENTATION, CONTROL, AND MODELING OF CONTINUOUS FLOW MICROWAVE PROCESSING

*Cristina Sabliov and Dorin Boldor*

## 7.1. Introductory comments

Microwave heating is emerging as a viable commercial alternative for high-temperature short-time processing of thermosensitive materials. As opposed to conventional heating methods, microwave heating takes place throughout the volume of the product. This volumetric heat delivery leads to a much higher rate of heating than conventional methods, limited by the heat penetration from the heated surface to the bulk of the material. Therefore, heating times are typically much shorter in microwave processing than in conventional heating processes. As a downside, the rapid heating may result in uneven temperatures in the heated material, which could negatively impact product quality. The uneven heat distribution specific to discontinuous microwave systems can be avoided by the use of continuous systems, where the electric field is focused and uniformly distributed in the heating cavity and throughout the product.

The heating of a dielectric material in the presence of an electromagnetic field is based on intermolecular friction that arises via ionic conduction and dipolar rotation (White, 1973; Schiffmann, 1997).

*Thermal Processing of Foods: Control and Automation*. Edited by K.P. Sandeep

The mechanisms of dielectric heating have been thoroughly analyzed and described in the literature (von Hippel, 1954; Rosenthal, 1992; Clark *et al.*, 1997; Schiffmann, 1997).

### *7.1.1. Advantages of continuous microwave systems over batch-type microwave systems*

Traditionally, microwave heating of foods has been limited to batch-type processes, which proved most efficient and economically feasible for quick heating and reheating of foods in small, home-type appliances. Industrial use of microwave heating in food processing has been hampered by the lack of technology that could generate a reliable uniform temperature distribution within the heated product. Therefore, industrial food applications have been generally limited to tempering of frozen meats prior to regular processing, or to large-scale batch-type applications. In these batch processes, the throughput is limited and the electric field distribution is nonuniform, similar to small home-type appliances (Rosenthal, 1992).

The nonuniformity of the electric field distribution due to multiple reflections of the wave within the microwave cavity leads to temperature nonuniformities within the heated product (i.e., hot spots). Thermal runaway effect (overcooking or burning) occurs in these hot spots, while other parts of the products remain underheated (Decareau, 1985). A rotating blade (or mode mixer) placed where the waves enter the cavity alleviates some of the problems associated with thermal runaway. Nevertheless, due to the nonuniformity of heating, most batch-type microwave processes are followed by additional unit operations to insure proper thermal treatment of the food product. By comparison, continuous microwave heating is performed in single-mode applicators that provide a uniform electrical field in the material undergoing heating.

### *7.1.2. Components of a continuous microwave system*

Any continuous microwave system designed for heating or drying applications is composed of a microwave generator and its control unit, a circulator, a water load, an applicator, a directional (or power) coupler, a tuning coupler, and the connecting waveguides. The

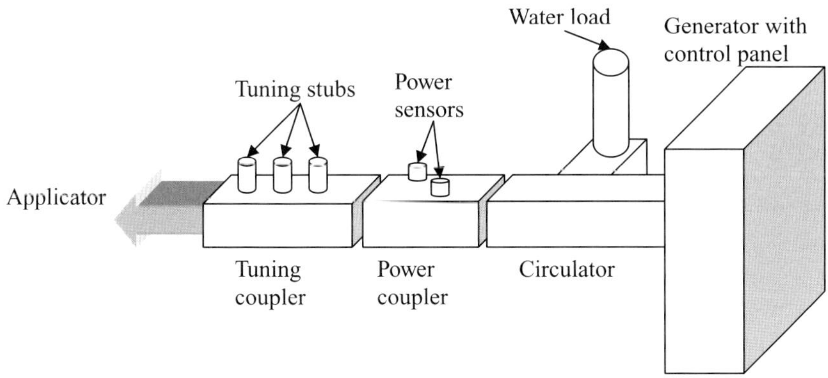


**Figure 7.1.** Components of a continuous microwave system.

generator contains a power supply for the magnetron, the magnetron that generates the microwave beam, the magnetron cooling system (with water), and the control panel from which the generator is operated (Figure 7.1).

Two types of applicators (geometrically different) are commonly used for: (1) continuous microwave heating of pumpable fluids and (2) heating/drying of solid foods (Roussy and Pearce, 1995; Metaxas and Meredith, 1983). Pumpable products are heated in a pipe made of a microwave transparent material, centered in a cylindrical-type focusing cavity. Microwaves delivered in a single mode (transverse electric— $TE_{10}$, or transverse magnetic—$TM_{01}$) to the cavity are reflected off the walls of the cavity and are focused in the center. This creates a uniform electric field at the center of the tube applicator through which the product is flowing. All cylindrical applicators have built-in safety devices at the location where the product tube is inserted into the focusing cavity, with the sole role of dissipating the microwave energy that otherwise would escape outside the cavity.

In drying applications, the most common type of applicator is a $TE_{10}$ traveling wave applicator, a modified waveguide allowing for the product to be conveyed on a belt placed at the center of the cavity (Jones, 1986; Metaxas and Meredith, 1983). In the center of a $TE_{10}$ waveguide, the electric field distribution is relatively uniform, and the dielectric material, running parallel with the electrical field is heated uniformly.

The traveling wave applicator can be manufactured either as a narrow waveguide, in which the electrical field decays exponentially from the entrance as microwaves are being absorbed and converted into heat, or as a wider system, in which the waveguide takes a serpentine design (Roussy and Pearce, 1995), and the electric field decays exponentially on each section of the applicator. The planar applicators also have built-in safety devices at the belt's entrance and exit to dissipate all microwave energy that otherwise might escape into the environment.

In between the generator and the applicator, a series of components are required for proper operation of the microwave system. Located close to the generator, a circulator acts as a gatekeeper to prevent reflected microwaves from returning to the generator and damaging the magnetron. The circulator is a modified waveguide, and it acts as a three-way valve (T-type) with a powerful magnet at the center. The generator and the applicator are connected to the ends of the "T", while the water load is connected at the bottom of the "T". The magnet in the center deflects the microwave radiation coming back from the applicator into the water load, preventing it from reentering the generator.

The power (or directional) coupler is usually installed between the circulator and the applicator, with two main purposes. The first purpose is to decouple the forward power and the reverse power such that the instrumentation system is able to differentiate between the power output of the generator (forward power) and the remaining reflected power from the applicator. The second purpose of the power coupler is to reduce the microwave power to a level measurable with various microwave sensors (discussed below), usually in the range of $-60$ to $-100$ dB.

The tuning coupler is a section of waveguide that has three tuning stubs, usually made of aluminum, inserted at specific distances and depths, depending on the microwave frequency and the dielectric properties of the product undergoing processing. The stubs act as mirrors that reflect the microwaves back into the applicator and increase the amount of energy being absorbed in the product. The tuning procedure (setting the depth of insertion) is performed using a network analyzer (Sucher and Fox, 1963; Roussy and Pearce, 1995).

All other connecting waveguides are made of aluminum at standard sizes.

## 7.2. The heating process in a continuous microwave system

Temperature increase of a product during microwave heating is caused by internal heat generation due to absorption of electrical energy from the microwave field. The heat generated is subsequently distributed throughout the product by conduction and convection (Rosenthal, 1992; Yakovlev, 2005). Temperature uniformity within a product heated by microwave energy is affected by the frequency, power, and geometry of the microwave heating system, the physical and electromagnetic properties of the flowing material, and the residence time of the flowing product through the microwave cavity. Mathematical description of microwave heating, a multiphysics phenomenon, requires coupling of the three components involved: electromagnetics, heat transfer, and fluid flow.

### *7.2.1. Governing equations*

The equations that describe the microwave heating process in a continuous system are summarized as follows. Maxwell's equations describe the electromagnetic field distribution in the microwave cavity.

$$\nabla \times H = J_s + j\omega\varepsilon_0[\varepsilon_r] \times E \tag{7.1}$$

$$\nabla \times E = -j\omega\mu_0[\mu_r] \times H \tag{7.2}$$

$$\nabla \times D = \rho \tag{7.3}$$

$$\nabla \times B = 0 \tag{7.4}$$

where H is the magnetic field intensity (A/m), $E$ is the electric field intensity (V/m), B is the magnetic flux intensity (Wb/m$^2$), D is the electric flux density (C/m$^2$), J is the electric current density (A/m$^2$), $\omega$ is the angular frequency (radian/s), $\varepsilon_0$ is the free space permittivity = $8.854 \times 10^{-12}$ F/m, $\varepsilon_r$ is the medium's relative permittivity, $\mu_0$ is the free space permeability = $4\pi \times 10^{-7}$ H/m, $\mu_r$ is the medium's relative permeability, $\rho$ is the electric charge density (C/m$^3$), and $j = \sqrt{-1}$.

The electric field intensity together with the conductivity of the material dictates the power loss in the dielectric:

$$Q = \frac{1}{2} \iiint_{v} \varpi \varepsilon'' |E|^2 \tag{7.5}$$

where $Q$ is the power loss (W/m$^3$), $\omega$ is the angular frequency (radian/s), $\varepsilon''$ is the dielectric loss (F/m), and $E$ is the electric field intensity (V/m).

The power loss coupled with the heat balance equation allows for the temperature of the microwaved product to be solved for:

$$\frac{\partial(\rho c_{\rm p} T)}{\partial t} = \nabla^2(kT) + Q \tag{7.6}$$

where $\rho$, $c_{\rm p}$, and $k$ are the density (kg/m$^3$), specific heat (J/kgK), and thermal conductivity (W/mK) of the dielectric, respectively, and $T$ is the temperature (°C), $t$ is the time (s), and $Q$ is the volumetric heat source (i.e., power loss) (W/m$^3$).

The velocity profile and residence time of the flowing product through the microwave cavity is provided by solving the fluid flow equation (Bird *et al.*, 1994):

$$\frac{D(\rho v)}{Dt} = -\nabla p + \rho g \tag{7.7}$$

where $\rho$ is the density (kg/m$^3$), $v$ is the velocity (m/s), $t$ is the time (s), $p$ is the pressure (N/m$^2$), and $g$ is the gravitational acceleration (9.8 m/s$^2$). It is important to note that the physical properties of the material, density, thermal conductivity, and specific heat are temperature dependent. Therefore, all three equations should be fully coupled to rigorously determine temperature change in a material undergoing continuous microwave heating.

### 7.2.2. Parameters that affect microwave heating

Parameters that critically affect the temperature increase and uniformity in a microwave-heated product are related to the product dielectric and thermal properties and to the design of the microwave system (i.e., power and frequency, geometry).

*7.2.2.1. Dielectric properties*

Dielectric properties of a material, among other parameters, dictate the amount of electrical energy absorbed by the material, and the extent to which this energy is converted into heat. Dielectric permittivity, $\varepsilon$, is defined as

$$\varepsilon = \varepsilon' - j\varepsilon'' \tag{7.8}$$

The real part of the dielectric permittivity ($\varepsilon'$) is called dielectric constant and it is an indicator of the amount of electrical energy absorbed by the material, whereas the imaginary part ($\varepsilon''$) of the dielectric permittivity, the dielectric loss, is an indicator of how much of this energy is converted into heat.

Dielectric properties of food materials depend upon the composition of the material, frequency, and temperature. Therefore, it is critical that the dielectric properties specific to a material at the working frequency and as a function of temperature be known prior to a microwave run. Foods at frequencies between 400 and 900 MHz show strong temperature-sensitive dielectric behavior, but the variation is less pronounced at 2800 MHz (Ohlsson and Bengtsson, 1975). Dielectric constant usually decreases with frequency (for frequencies between the standard 915 and 2450 MHz) and it increases with temperature (over 0°C). Generally, dielectric loss of materials decreases with frequency as well, but it may increase or decrease with temperature at a specific frequency. Changes in dielectric properties alter the penetration of radiation during heating, and a sharp increase in the dielectric loss with temperature can lead to runaway conditions (Ayappa *et al.*, 1991).

Some materials have low dielectric properties, which translate into a limited ability to absorb and convert electric energy into heat; these materials are poor candidates for microwave heating. Microwave heating of other materials such as water, with a high dielectric constant (79 at 25°C and 915 MHz) and dielectric loss (3.7 at 25°C and 915 MHz), is very efficient. The rule of thumb is that materials with high moisture contents or high ionic concentrations (i.e., salt) heat up quickly by microwave heating.

The power loss in the dielectric (equation (7.5)) is directly proportional to dielectric loss and the electric field intensity. It follows that the higher the dielectric loss (and the higher the electric field intensity), the higher the power loss is, and consequently the higher the temperature of the material. The depth of penetration, defined as the distance from the surface into the material at which the power drops to $e^{-1}$ of the original value (Metaxas and Meredith, 1983) is also related to the dielectric properties of the material, as follows:

$$D_p = \frac{\lambda_0}{2\pi(2\varepsilon')^{1/2}}\left[\left(1+\left(\frac{\varepsilon''_{\mathrm{eff}}}{\varepsilon'}\right)^2\right)^{1/2}-1\right]^{-1/2} \qquad (7.9)$$

where $\lambda_0$ is the free space wavelength of the microwave radiation ($m$). The penetration depth of microwave energy is smaller for those dielectrics characterized by a higher loss factor, defined as the ratio between the dielectric loss and the dielectric constant. When the penetration depth is larger than the sample, the temperature increase due to microwave energy is more uniform throughout the sample. In contrast, when the penetration depth is a significantly smaller than the size of the sample, heating will take place at the surface of the food (similar to traditional heat transfer), and it will result in nonuniform temperature distribution.

#### *7.2.2.2. Microwave power and frequency and geometry of the system*

The heat loss in the dielectric is proportional to the square of the electric field, which increases with an increase in the incident power. It follows that the higher the incident power level, the increased electric field intensity, and the higher the heat loss in the dielectric. It also follows that a nonuniform electric field density in the dielectric can be responsible for nonuniform volumetric heating of the material, and consequently a nonuniform temperature in the heated material. If the electric field intensity is dictated by the microwave power, electric field distribution is controlled by the design of the microwave cavity. A microwave system can be specifically designed to efficiently and uniformly heat a material of known dielectric properties, by selectively choosing the microwave system characteristics, such as power and geometry.

Industrial microwave frequencies are limited to 915 and 2450 MHz in the United States (Roussy and Pearce, 1995). Penetration depth (equation (7.9)) is different for the two frequencies, not only because dielectric properties change as a function of frequency but also due to the decrease in the wavelength with frequency, as follows:

$$\lambda_0 = \frac{c}{f} \tag{7.10}$$

where $\lambda_0$ is the free space wavelength of the microwave radiation ($m$), $c$ is the speed of light ($c = 3 \times 10^8$ m/s), and $f$ is the frequency (Hz). Equations (7.9) and (7.10) show that as frequency increases, the wavelength and the penetration depth decrease.

## 7.3. Instrumentation of continuous flow microwave systems

Typical instrumentation for continuous flow microwave heating includes a set of sensors (power, temperature, and flow rate, and in the case of drying, moisture), a data acquisition system and a recording unit. Data recording can be performed using analog (charts) or digital devices (programmable logic controllers (PLCs) or a data acquisition board connected to a computer).

### *7.3.1. Power and temperature sensors*

In continuous microwave processing of materials, the most important parameters that need to be measured, monitored, and controlled are microwave power level, temperature of the product throughout or at the end of the process, and for drying, the moisture content of the product.

#### *7.3.1.1. Power sensors*

The power level is usually the critical factor in the design and the purchasing decisions involving continuous microwave heating systems. Power measurement is also critical in determining the amount of microwave energy available for heating and the actual amount of energy converted into heat during processing.

Power sensors need to be selected based on the requirement of the particular application, their sensitivity, dynamic range, and maximum power that can be measured with these sensors. Three popular sensors are currently in use for microwave power measurements: thermistors, thermocouples, and diode detectors (Roussy and Pearce, 1995; Agilent Technologies, 2001). An alternative method to measure the power is the calorimetric method, described later.

Thermistors are devices that translate the microwave energy into a change in their electrical resistance measured using electronic instrumentation. Thermocouples measure temperature increase induced in the junction exposed to microwave energy. The temperature increase is translated into a measurable voltage through the Seeback effect. Thermistors and thermocouples are always used in conjunction with a directional coupler that may have a reducing power factor of more than -–60 dB due to the low power (<10 W) measured by these devices (and power diodes, discussed below). The detailed principles of operation for thermistors and thermocouples in microwave applications are described in detail in Roussy and Pearce (1995) and Agilent Technologies (2001).

Power diodes convert the equivalent AC signal of a microwave beam to a DC voltage signal proportional to the microwave power. This signal can be measured using the diode calibration curves and a microwave power meter or a data acquisition system. Diodes are more sensitive than thermocouples and thermistors, and their slower response time is not a major disadvantage in microwave processing (as opposed to wireless communication, where speed is also important). Therefore, power diodes are the most common devices for power measurement in a continuous microwave processing unit.

An alternative to power sensors in measuring microwave power is the calorimetric technique. In this method, a known material (usually water) flows in a tube placed at an angle (usually 45°) across the waveguide. If the flow rate of the material and the inlet and outlet temperatures are known, the power of the microwave beam can be calculated using material and energy balance equations:

$$P = m\, c_p \Delta T \tag{7.11}$$

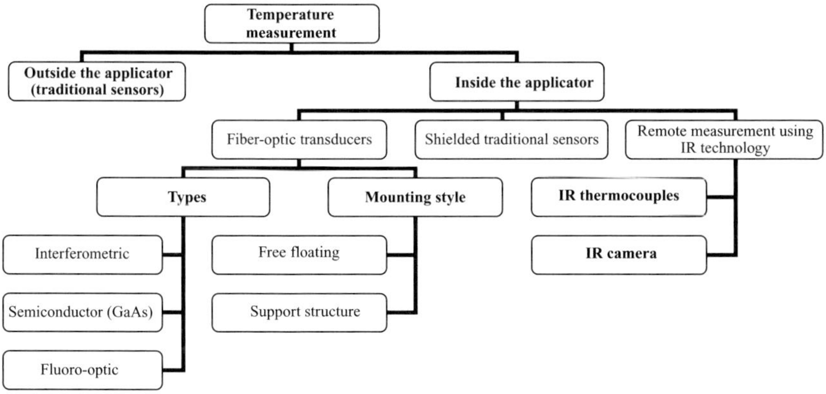


**Figure 7.2.** Temperature measurement options in microwave processing.

Calorimetric measurements can be very precise and can be used to measure the local volumetric power distribution and the magnitude of the electric field in a microwave cavity. However, calorimetric methods are limited to multi-directional power measurement, which is a disadvantage.

### *7.3.1.2. Temperature sensors*

There are two general approaches to measure temperature in microwave applications, each of them having certain advantages and disadvantages (Figure 7.2). The first approach is to measure the temperature of the food immediately outside the microwave applicator cavity using traditional measurement systems (thermistors, thermocouples, etc.). The second approach is to measure the temperature distribution inside the microwave applicator cavity using either innovative designs of the traditional sensors, which account for the peculiarities of the microwave processing, or nontraditional measurement systems such as fiber-optic temperature probes and remote measurements of surface temperature distributions using infrared technology (Goedeken *et al.*, 1991; Mullin and Bows, 1993).

The temperature of pumpable products flowing through a single-mode cavity (i.e., $TE_{10}$ or $TM_{10}$) is measured outside the microwave cavity using a single sensor or a profile probe for multiple measurements placed at the exit of the applicator. The type of sensor used (thermocouple, thermistor, etc.) is mostly dependent on the specific properties of the product undergoing microwave processing and the temperature requirements of the process.

These traditional temperature sensors cannot be used for measurements inside the microwave applicators due to the nature of microwaves and their interaction with metals. The presence of metallic components in a waveguide would change the distribution of microwaves in an applicator, a critical parameter for optimum product safety and quality. Metallic sensors, unless heavily shielded, are also prone to generation of sparks, which present a safety hazard and usually render the sensor useless.

Fiber-optic temperature probes, immune to electromagnetic radiation, are usually used in measuring the temperature of the food inside the microwave cavity. For batch processing, the fiber-optic instrumentation system can be easily designed to measure the temperature inside the food product at a predetermined location. This approach can be used both in large-scale batch processing applications and in laboratory-scale applications (Mullin and Bows, 1993). The fiber-optic probes can also be used (with some limitations) in continuous microwave processing of fluids in cylindrical applicators. They have to be either free floating in the product stream or anchored on a specially designed frame. The free floating fiber-optic probes, aligned at the center of the tube in the fluid at the highest velocity, will allow for the measurement of the temperature distribution along the center of the fluid column. Alternatively, a fiber-optic support system can be designed inside the pipe at predetermined positions, but this design may increase the cost significantly and may also change the flow distribution of the product.

Fiber-optic temperature probes come in three different types (Figure 7.2). The first design is based on the Fabry–Perot interferometer principle, in which the sensing tip of the probe is composed of a glass cavity of very precise dimensions (Lee and Taylor, 1991; Berghmans *et al.*, 1998; Degamber and Fernando, 2003). The measurement of the cross-correlation between the light emitted by the signal conditioner and the reflected beam allow the accurate determination of the sensing cavity dimensions. As the temperature around the probe tip changes, the dimensions of the cavity change, and therefore the temperature can be accurately measured. A second type of fiber-optic temperature sensor is based on semiconductor technology, in which the sensor at the tip of the probe is made out of a material (i.e., GaAs) with temperature dependent light-absorption properties (Berghmans *et al.*, 1998). The intensity of light reflected from the end of the tip is compared with the light emitted by the signal conditioner, which then processes the signal to give an accurate temperature measurement. The fluoro-optic probes, third option, contain a fluorescent material which starts emitting light when activated by an optical signal generated by the signal conditioner (Lyons *et al.*, 1985; Chan *et al.*, 1988; Sun *et al.*, 1989; Berghmans *et al.*, 1998; Reid *et al.*, 2001). The emitted light decay, and the decay time, dependent on temperature, is measured by the signal conditioner.

Another approach for measuring the temperature distribution is by using remote temperature measurement using infrared technology. For pumpable products, the surface temperature of the processing tube inside the microwave applicator is measured with an infrared camera (Goedeken *et al.*, 1991) through "pinholes" in the focusing cavity wall, which allow passing of infrared radiation while blocking microwave frequencies. Transport equations are afterwards used to infer the internal temperature distribution of the product inside the tube. In continuous microwave drying of solid materials in traveling wave applicators, the infrared technology can also be used to measure the surface temperature distribution of the dielectric and to relate it to the internal temperature (Boldor *et al.*, 2005). For drying applications, the measurement can be performed either with infrared thermocouples (i.e., model OS36-T, OMEGA Engineering, Inc., Stamford, CN) inserted at certain locations along the microwave waveguide (Boldor *et al.*, 2004) or with an infrared imaging system (Goedeken *et al.*, 1991).

### *7.3.2. Data acquisition*

The regulatory agencies require proper documentation and record keeping for all the parameters of interest in microwave processing, as with any other thermal processing. For microwave processing the parameters of interest, which must be monitored and recorded, are the temperature and microwave power levels. Data acquisition (i.e., temperature, power) can be performed digitally (computer based), or by use of traditional analog data recorders (charts). The analog data recorders have been widely used in the industry and are not discussed further in this chapter. Computer-based devices are emerging as a reliable alternative to traditional analog devices. Computer-based data acquisition systems allow for more complex information to be gathered about a specific process and for additional processing of information. An example of computer-based data acquisition system is the HP34970A (Agilent, Palo Alto, CA), or the DAQPad-6015 (National Instruments Corp., Austin, TX) connected to a computer controlled by a master software routine written in LabVIEW (National Instruments Corp., Austin, TX) on a Microsoft Windows NT (Microsoft Corp., Redmond, WS) operating system. The same software monitors and records the temperature, the power levels, and the other parameters of interest in the microwave applicator. Computer-based data acquisition and control systems can be as reliable as the operating system they run on. There is a trend towards moving computer-based data acquisition towards operating systems that are independent of general software support platforms such as Microsoft Windows. For example, National Instrument's LabVIEW software can be easily developed on Microsoft Windows or other operating systems, and then ported to an independent platform running on a more reliable LabVIEW Real-Time operating system. These platforms allow integration of more advanced monitoring capabilities such as imaging systems, Internet and wireless communications (which are being integrated even in lower level devices) with the traditional sensors. PLCs are also reliable digital recorders that do not require a supporting operating system. They are fairly easy to program and cost less than their more complex computer-based counterparts. One disadvantage is that the simplicity of PLCs does not allow integration of advanced processing capabilities (i.e., imaging).

## 7.4. Control of continuous flow microwave processing system

### *7.4.1. Process control theory*

Industrial processes are not ideal; they are subject to external variables (i.e., environmental variables) and disturbances such as changes in feed compositions. In addition, they may show unsteady-state behavior, especially at the start or at the end of the process. Therefore, the control of unit operations is critical to reach and maintain the processing operation in a safe region yielding consistent results. The most important and widely used process control method is feedback control (Marlin, 2000). Other control methods include feedforward control, combinations of feedback/feedforward, and more advanced control methods that are beyond the scope of this chapter (such as model predictive and inferential control) and are not discussed here. For a more in-depth analysis of the various control methods, the reader is directed to various books on this topic referenced at the end of this chapter (Ogunnaike and Ray, 1994; McFarlane, 1995; Mittal, 1996; Marlin, 2000; Khalil, 2002; Seborg *et al.*, 2003; Morari *et al.*, 2005).

In general, all dynamic processes, including microwave heating, can be characterized by their transient behavior as either a first- or a second-order systems, and usually the analog processes are represented in the Laplace domain (digital processes are represented in the $z$-domain). In these systems, the manipulation of an input variable, such as the microwave power level will induce a response in a measured variable, such as temperature. The response of the process is proportional with the process gain ($K_p$), and it is also characterized by the process time constant, dead time, and in the case of second-order systems by the damping coefficient. The process representation in the Laplace domain is defined by the following transfer functions (Marlin, 2000):

$$\text{First order: } G_p(s) = \frac{Y(s)}{X(s)} = \frac{K_p e^{-\theta s}}{\tau_p s + 1}$$

$$\text{Second order: } G_p(s) = \frac{Y(s)}{X(s)} = \frac{K_p e^{-\theta s}}{\tau_p^2 s^2 + 2\xi \tau_p s + 1}$$

where $G_p(s)$ is the transfer function of the process (Laplace domain), $X(s)$ is the input function (Laplace domain), $Y(s)$ is the output function (Laplace domain), $K_p$ is the process gain, the ratio of the final steady-state value to the initial steady-state value, $\theta$ is the dead time of the process, $\tau_p$ is the time constant of the process, $s$ is the Laplace domain variable, and $\xi$ is the damping coefficient.

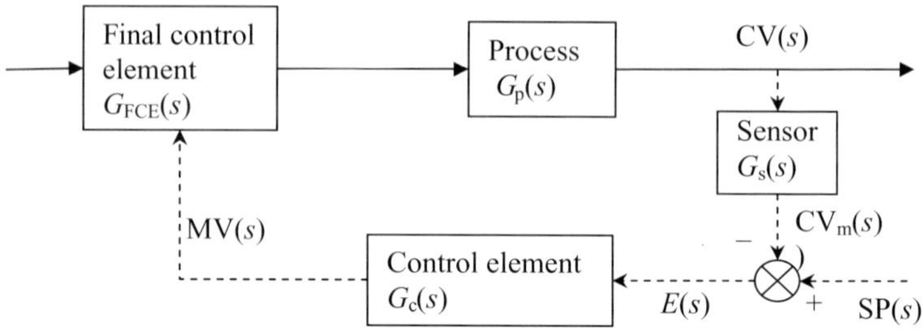


**Figure 7.3.** Schematic of the feedback control loop. CV($s$), controlled variable (Laplace domain, time domain); $CV_m(s)$, measured controlled variable (Laplace domain); E($s$), error (Laplace domain); $G_c(s)$, transfer function of the controller (Laplace domain); $G_{FCE}(s)$, transfer function of the final control element (Laplace domain); $G_s(s)$, transfer function of the sensor (Laplace domain); MV($s$), manipulated variable (Laplace domain); SP($s$), set-point (Laplace domain, time domain).

### *7.4.2. Feed back control*

The microwave heating processes (for both resonant cavity and traveling wave applicator) can be modeled as a first-order process with dead time, and it can be controlled via several methods. The simplest control is the feedback-loop method, shown in Figure 7.3. The most important characteristics of this method are its simplicity and its property of not requiring any *a priori* information about the process.

The process parameters ($K_p$, $\theta$, and $\tau_p$) can be established using process reaction curves (Figure 7.4) as described in literature (Marlin, 2000). A step change in the manipulated variable (microwave power) induces a first-order response in the measured variable (temperature). The major advantage of this method is that the transfer functions of the sensor $G_s$ and of the final control element $G_{FCE}$ are included in the model.

The feedback controller reacts to deviations from a set point in the measured variable (i.e., temperature in microwave processing) by adjusting the manipulated variable (microwave power) using a proportional, integral, and derivative (PID) algorithm. The mathematical representation of the PID controller in the Laplace domain is shown in Equation 3:

$$G_c(s) = K_c\left(1 + \frac{1}{\tau_i s + 1} + \tau_d s\right) \qquad (7.12)$$

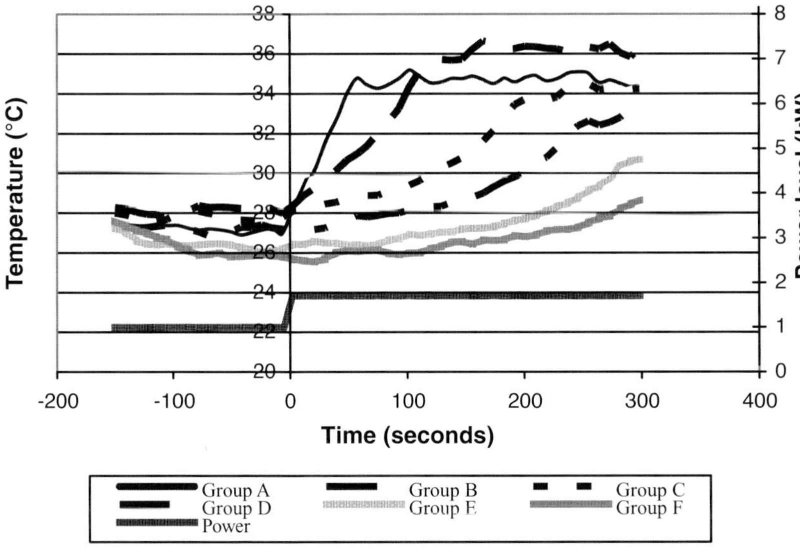


**Figure 7.4.** Process reaction curves in continuous flow microwave drying. Infrared temperature sensors placed closer to the microwave source (A, B) have better response time than those placed further away (C–F). Reproduced from Boldor *et al.* (2004).

where $K_c$ is the proportionality constant, $\tau_i$ is the integral time, and $\tau_d$ is the derivative time.

The PID feedback controller requires a tuning procedure, in which the optimum operating parameters of the controller ($K_p$, $\tau_i$, and $\tau_d$) are determined. An initial set of parameters is chosen based either on methods existent in literature (such as the Ciancone open-loop method, Ciancone and Marlin, 1992) or on software simulations of the microwave heating process (discussed below). Once the parameters are established, the control algorithm can be tested using a servo scenario (set-point tracking), in which the set-point is changed periodically and the system is allowed to reach steady-state operation after each change. An example of feedback control using the servo scenario in continuous flow microwave drying is shown in Figure 7.5.

The feedback control is a powerful and reliable control method, but sometimes it has to be coupled with other control methods (i.e., feedforward). The feedforward control is mostly applied when a feedforward variable, independent of other variables that may

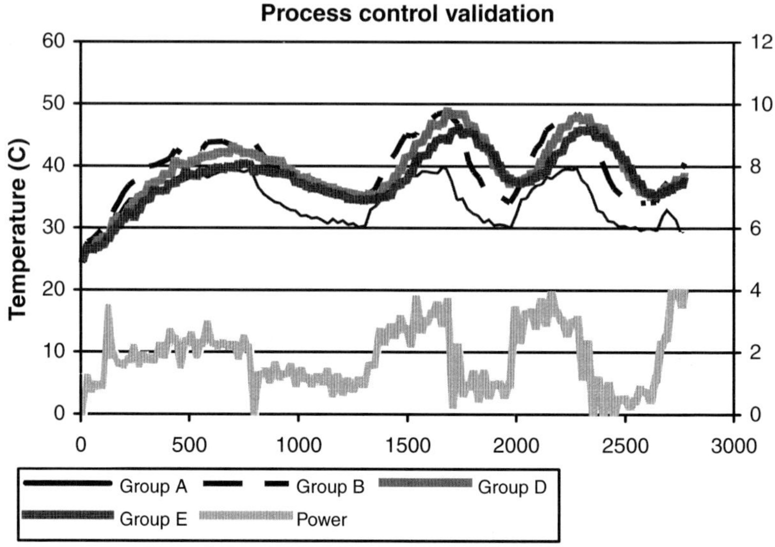


**Figure 7.5.** Set-point tracking in continuous flow microwave drying using PI feedback control. The controller is adjusting the power level to match certain desired temperatures (30 and 40°C for the A group of sensors). The time required to reach stable operation is around 400 seconds. Reproduced from Boldor *et al.* (2004).

affect the process, is available. Another required condition is that the feedforward variable must indicate the occurrence of a significant disturbance. In addition, when used in combination with the feedback controller, the dynamics of the disturbance must be comparable (or slower than) to the dynamics of the output variable manipulated by the feedback controller (in other words, there is a long time between the disturbance and the correction made by the feedback controller). If these conditions are not met, using the feedforward control method can lead to unstable process operation.

### *7.4.3. Process control applications: PLCs and software-based control*

Software-based control methods rely mainly on computer programs that can interpret the data collected by the sensors and take the appropriate actions. The major advantage of these systems is their scalability, the ease of implementing complex control algorithms, and the virtually unlimited number and types of sensors that can be used. While these programs are very cumbersome to code, new programming languages, such as graphical programming *G*, or Matlab's Simulink (Mathworks, Inc., Natick, MA) and their seamless integration with new hardware allow rapid development of a variety of control strategies and data acquisition systems. One of the major advantages of using software-based control is the possibility to integrate imaging and computer vision systems and algorithms not possible by using simpler methods such as PLCs. For example, in continuous flow microwave drying, infrared imaging can be integrated with other sensors on the same software routine written in LabVIEW to control the microwave generator based on the spatial and temporal temperature distribution, which cannot be measured with other sensors.

Another powerful feature of software based methods are their capability to simulate a virtually infinite number of processing disturbances and processing cases, allowing the development of control strategies for each and every one of them (Figure 7.6). The simulations reduce the costs of development and deployment of control systems, and allow simulations of the processes behavior in the presence of the less complex PLCs.

Computer-based control systems have the same disadvantage as computer-based data acquisition systems, both relying on a third-party operating system to function properly and being rather expensive. For simple control systems, which do not require unusual sensors such as imaging devices, the PLCs present a cheaper and more reliable alternative, since they are smaller, do not require an operating system, and the newer versions have enhanced features such as autotuning and computer and Internet connectivity.

## 7.5. Modeling of continuous flow microwave systems

### *7.5.1. General information*

Temperature uniformity is an indicator of the quality of the microwave process and it can be determined experimentally for a

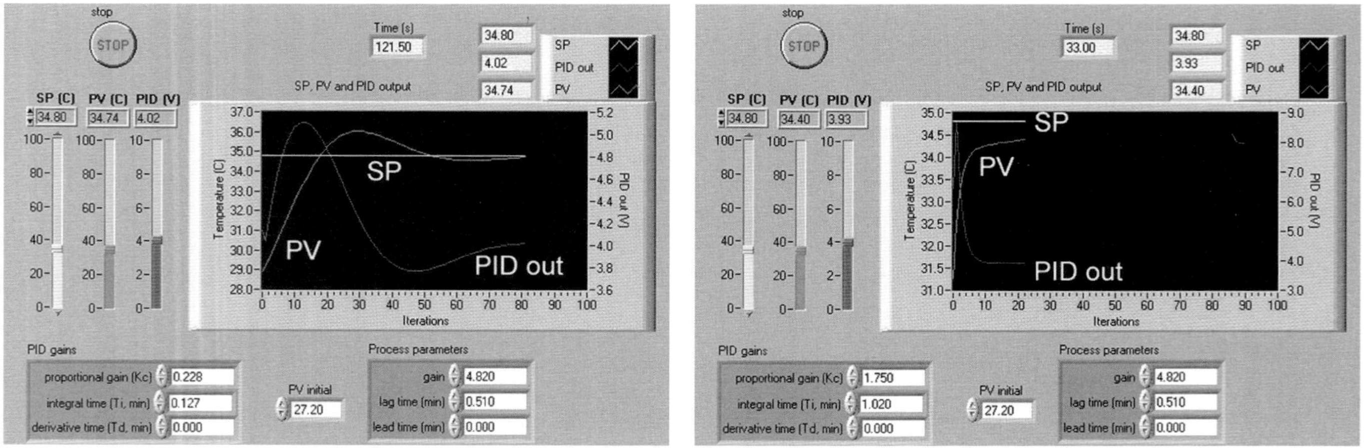


**Figure 7.6.** Simulation results for control of continuous flow microwave process for initial (left) and optimum (right) tuning parameters. The simulation was performed using LabVIEW. Reproduced from Boldor *et al.* (2004).

combination of parameters (product dielectric properties, cavity geometry and dimensions, and applicator diameter), or it can be computed numerically. Numerical methods are preferred to experimental methods when the influence of a wide array of parameters on the temperature profile is of interest. Modeling does not replace experiments, however; the model needs to be validated against experimental data for a set of conditions. Following validation, the model can be extensively used to simulate the microwave heating of a product under a combination of parameters, which greatly aids in understanding of the microwave heating process in its complexity.

As mentioned earlier, temperature uniformity within a product heated by microwave energy is affected by the design of the continuous microwave system, the physical and the electromagnetic properties of the flowing material, and the residence time of the flowing product through the microwave cavity. Mathematical description of microwave heating, a multiphysics phenomena, requires coupling all three components involved, electromagnetics, heat transfer, and fluid flow.

Due to the complexity of the process, currently available numeric models only partially address the three mentioned components. Most literature reports involve solving Maxwell's equations and computing the electric field intensities. The techniques can be broadly classified into methods using either an integral or a differential formulation of Maxwell's equations (Speigel, 1984). Differential forms of Maxwell's equations are solved either in the time or in the frequency domain. In the latter method, the time derivatives are eliminated by assuming that the field variables have a frequency dependence of $e^{j\omega t}$ (Balanis, 1989). In early published work on microwave heating simulation, the electromagnetic field distribution in multimode microwave ovens was numerically determined and the electric field distribution was used as an indicator of the heating uniformity (Jia and Jolly, 1992; Ohlsson, 1993). Other models consisted of electromagnetic field analysis coupled with heat transfer phenomena via the microwave heat generation term (Taflove and Brodwin, 1975; Nachman and Turgeon, 1984; Chen *et al.*, 1993; Lin *et al.*, 1995; Vilayannur *et al.*, 1998; Alpert and Jerby, 1999; Sabliov *et al.*, 2004). Models for predicting the temperature distributions in microwave heating processes accounting for temperature dependent dielectric and thermophysical properties of the product were also developed (Ayappa *et al.*, 1992; Pichon *et al.*, 1994; Sekkak *et al.*, 1994; Barrat *et al.*, 1995; Kanellopoulos *et al.*, 1995; Devece *et al.*, 1998; Soriano *et al.*, 1998).

Attempts to solve for the temperature profile in a liquid flowing through a microwave cavity were also made. Mudgett (1985) developed a model for a continuous-flow microwave system by using the plug flow approach and assuming a homogenous heat dissipation of the microwave energy. Datta and Liu (1992) used a heat generation term that exponentially decayed from the surface into the liquid to solve the energy and flow equations to predict temperature inside a cylindrical container of liquid in a microwave cavity. LeBail *et al.* (2000) developed a model to predict the temperature profiles under continuous tube-flow microwave heating by assuming a uniform volumetric power in the core of the flow. Assumptions were made in all previous studies regarding the power distribution inside the loaded cavity, assumed either constant power or power undergoing exponential decay.

Several software packages are available to simulate the process of microwave heating, some based on the FDTD, others on FEM, as compiled in Table 7.1. Priced differently, with different complexities and features, these software packages may offer advantages specific to the type of food application sought after. Out of the listed packages, ANSYS (ANSYS, Inc., Pittsburgh, PA) was found most capable of handling multiphysics problems. A case study is presented below, showing the possibility offered by ANSYS to couple high-frequency electromagnetics with heat transfer and fluid flow in an attempt to solve for the temperature profile in a product undergoing continuous microwave heating.

### *7.5.2. Case study: heating of milk in a 5 kW continuous microwave system at 915 MHz*

In order to simulate the temperature in a product subject to high-frequency fields, electric field distribution in the microwave cavity must be first computed, then used to determine the power loss, as described by equations (7.5) and (7.6). Next, equations (7.7) and (7.8) are numerically solved to find the temperature and the velocity of the heated material. Information on the dielectric properties of

**Table 7.1.** Available modeling software packages most suitable for modeling systems and components of MW power engineering

| EM | Coupling | Code | Vendor |
|---|---|---|---|
| | | *HFSS* | Ansoft, Inc. (www.ansoft.com) |
| | √ | *Multiphysics* | ANSYS (www.ansys.com) |
| | √ | *FEMLAB* | COMSOL (www.comol.com) |
| √ | | *Microwave Studio* | CST (www.cst.com) |
| √ | | *MEFiSTo* | Faustus Scientific Corp.(www.faustcorp.com) |
| | | *Micro-Strips* | Flomerics Ltd. (www.micro-strips.com) |
| √ | (√) | *QuickWave-3D [a])* | QWED Sp. z o.o. (www.qwed.com.pl) |
| | | *FIDELITY* | Zeland Software Inc. (www.zeland.com) |

*Source*: Yakovlev, 2005.
[a]Also distributed under the name Concerto by Vector Fields, Ltd (www.vectorfields.com).

product, power, frequency and geometry of the system, and boundary conditions must be specified.

For the specific case study, the analysis was performed in ANSYS (ANSYS, Inc., Pittsburgh, PA) with milk as the product of interest. High-frequency electromagnetics and Flotran were iteratively coupled via the heat generation term and solved for a steady-state situation. The temperature and velocity profiles were traced in milk heated in a 5kW IMS (Industrial Microwave Systems, Morrisville, NC) operating at 915 MHz.

The ANSYS model consisted of three steps, preprocessing, processing (solving) and postprocessing. Preprocessing consisted of geometry and mesh generation and specification of the initial and boundary conditions, and material properties. Processing involved solving Maxwell's equation, heat and flow equations under the specified boundary conditions. Postprocessing was used to visualize, plot, and analyze results. The steps involved are described in more detail, as follows.

**Table 7.2.** Material dielectric properties

| Product | Dielectric constant ($\varepsilon'$) | Dielectric loss ($\varepsilon''$) | Loss tan ($\tan \delta$) |
|---|---|---|---|
| Skim milk | 64.6 | 19.58 | 0.31 |

*7.5.2.1. Preprocessing*

7.5.2.1.1. Geometry and grid generation

The geometry was created based on a continuous microwave system built by Industrial Microwave Systems (IMS, Inc., Morrisville, NC). The system consisted of the waveguide, the elliptical focusing cavity, and the cylindrical applicator. Next, the geometry was meshed with a tetrahedral element grid (HF119), coarser in the waveguide and elliptical cavity and finer in the applicator tube. For the Flotran analysis, a FLUID142 element was applied to the applicator tube to model the steady-state fluid flow/thermal phenomena. Within the tube, a higher density grid was applied near the walls since higher heating rates as well as a velocity boundary layer was expected at this location. A MESH200, a "mesh-only" element was applied to the waveguide for the Flotran analysis, where the heat and momentum equations were not solved.

7.5.2.1.2. Material properties

Skim milk was selected as the material of choice. Dielectric properties are listed in Table 7.2. The density, viscosity, thermal conductivity, and specific heat were all assumed constant with temperature and equal to 1030 kg/m$^3$, $10^{-3}$ kg/ms, 0.61 W/mK, and 3965 J/kgK, respectively.

7.5.2.1.3. Boundary conditions

*High-frequency electromagnetics*: Maxwell's equations were solved under the following assumptions. Energy (5 kW) was supplied to the cavity by means of the $TE_{10}$ mode, the dominant mode in a rectangular waveguide, at a frequency of 915 MHz. A rectangular high-frequency port was set at the waveguide interfacing the generator; the port was designed such as to appear transparent to the energy reflected back due to a mismatch between the impedance of the waveguide and that of the dielectric in the cavity. The walls of the waveguide were assumed perfectly conductive. The applicator tube was assumed negligible in thickness and transparent to microwaves.

*Flotran*: The flow was assumed laminar, with an initial velocity corresponding to a flow rate of 2 l/min. Zero velocity was applied at the wall. The inlet temperature was set to 25°C, and the outlet pressure was set to atmospheric pressure. Adiabatic conditions were applied at the applicator wall (no heat was assumed exchanged between the dielectric and the air in the elliptical cavity).

*7.5.2.2. Processing*

First, the electromagnetic algorithm was run using the specified material dielectric properties at the working frequency of 915 MHz. The dissipated power at each cell in the domain was calculated from the rms field values and the properties of the material. Next, the dissipated power was uploaded into the Flotran module, and the velocity and temperature solved for, by solving the momentum and the heat equations. A segregated sequential solver algorithm was used. The flow equation was solved separately from the heat equation. The sequence could be repeated to account for the change in the dielectric and physical properties of the material as a function of temperature.

*7.5.2.3. Postprocessing*

Temperature and velocity profiles were traced as a function of space and time, using the postprocessing tool. The results are presented in Figure 7.7.

## Conclusions

Continuous processing of food products using microwave heating is emerging as a promising alternative for current heating technologies. The fast heating rates achieved in microwave heating allow for increased production speeds, eliminating the process dead time and significantly reducing the time constant of the process when compared to other heating processes. The real-time measurement of

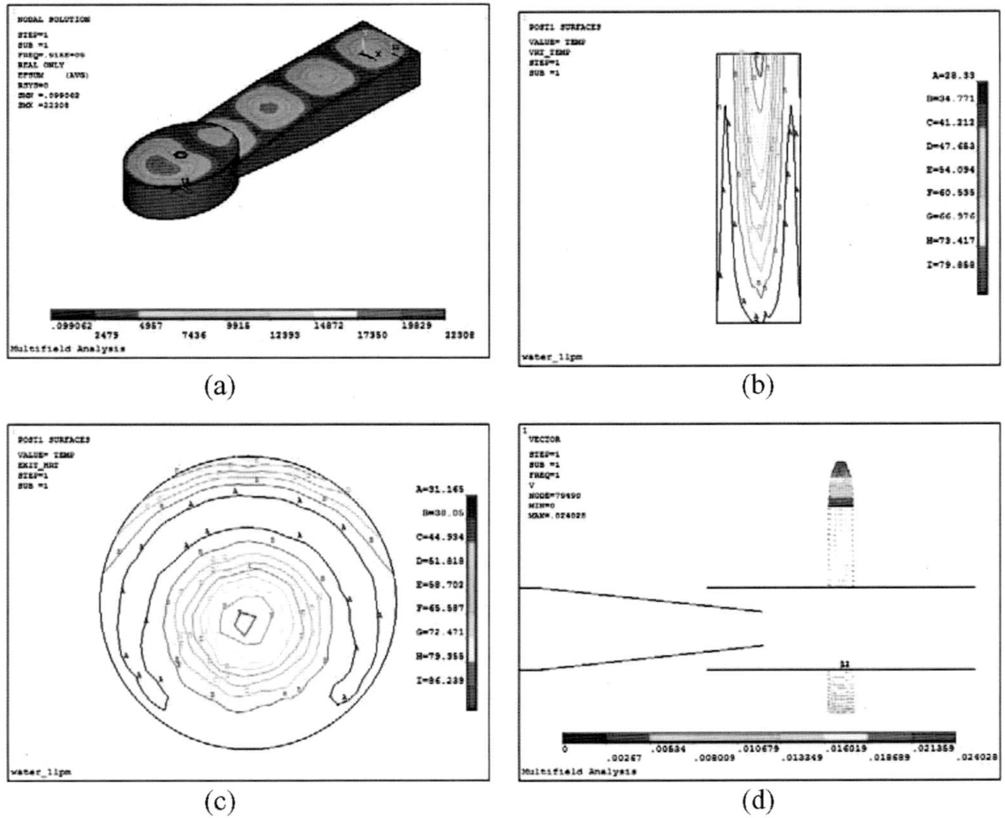


**Figure 7.7.** (a) Electric field distribution in the waveguide and cavity. (b) Temperature distribution in the cross-axial planar section ($x$–$y$ plane) of the material in the applicator. (c) Spatial temperature distribution ($x$–$z$ plane) of the material in the applicator. (d) Velocity profile of the flowing material.

process parameters (power, temperature) allow for the control of the microwave processing to be performed instantaneously. Temperature measurement inside the microwave cavity during the heating process requires the use of innovative sensors and technologies, which may increase the up-front cost of the instrumentation, but delivers information otherwise unattainable. Modeling of the continuous flow microwave processing provides a powerful tool for food engineers and equipment manufacturers to simulate any number of processing conditions and use the generated information to optimize the microwave heating process.

## References

Agilent Technologies. 2001. Fundamentals of RF and microwave power measurements. Application note 64-1C. Doc. No. 5965-6630E. USA.

Alpert, Y. and Jerby, E. 1999. Coupled thermal-electromagnetic model for microwave heating of temperature-dependent dielectric media. *IEEE Transactions on Plasma Science*, 27(2): 555–562.

Ayappa, K.G., Davis, H.T., Davis, E.A., and Gordon, J. 1991. Analysis of microwave heating of materials with temperature dependent properties. *AlchE Journal*, 37: 313.

Ayappa, K.G., Davis, H.T., Davis, E.A., and Gordon, J. 1992. Two-dimensional finite element analysis of microwave heating. *AIChE Journal*, 38(10): 1577–1592.

Balanis, C.A. 1989. *Advanced Engineering Electromagnetics*, Wiley, New York.

Barrat, L., Bows, J., Ma, L., Mullin, J., Paul, D., Pothecary, N., Railton, C., and Simons, D. 1995. Experimental validation of a combined electromagnetic and thermal FDTD model of a microwave heating process. *IEEE Transactions of Microwave Theory and Techniques*, 43(11): 2535–2572.

Berghmans, F., Vos, F., and Decréton, M. 1998. Evaluation of three different optical fibre temperature sensor types for application in gamma-radiation environment. *IEEE Transactions on Nuclear Science*, 45(3): 1537–1542.

Bird, R.B., Stewart, W.E., and E.N. Lightfoot. 1994. *Transport Phenomena*, John Wiley & Sons, New York.

Boldor, D., Sanders, T.H., and Hale, S.A. 2004. Control of continuous microwave drying process of peanuts using remote temperature measurement. In: *4th World Congress on Microwave & RF Applications Proceedings CD ROM*, Austin, Texas, November 2004.

Boldor, D., T.H. Sanders, K.R. Swartzel, and Farkas, B.E. 2005. A model for temperature and moisture distribution during continuous microwave drying. *Journal of Food Process Engineering*, 28(1): 68–87.

Chan K.W., Chou C.K., McDougall J. A., and Luk K.H. 1988 Changes in heating patterns due to perturbations by thermometer probes at 915 and 434 MHz. *International Journal of Hyperthermia*, 4: 447–56.

Chen D.S., Singh, R.K., Haghighi, K., and Nelson, P.E. 1993. Finite element analysis of temperature distribution in microwaved cylindrical potato tissue. *Journal of Food Engineering*, 18: 351–368.

Ciancone, R. and Marlin, T. 1992. Tune controllers to meet plant objectives. *Control*, 5: 50–57.

Clark, D.E., Sutton, W.H., and Lewis, D.A. 1997. Microwave processing of materials. In: *Microwaves: Theory and application in materials processing IV*, D.E., Clark, W.H., Sutton, D.A., Lewis (eds.). The American Ceramic Society, Westerville, OH.

Datta, A.K. and Liu, J. 1992. Thermal time distributions for microwave and conventional heating of food. *Transactions of IchemE, Part C*. 70: 83–90.

Decareau. 1985. Microwave in the food processing industry. Academic Press, Inc., New York.

Degamber, B. and Fernando, G.F. 2003. Microwave processing of thermosets: non-contact cure monitoring and fibre optic temperature sensors. *Plastics, Rubber and Composites*, 32(8/9): 327–333.

Devece, D., de los Reyes, E., and Soriano, V. 1998. A finite element and finite difference formulations for microwave heating laminar materials. *Journal of Microwave Power and Electromagnetic Energy*, 33(2): 67–76.

Goedeken, D.L., Tong, C.H., and Lentz, R.R. 1991. Design and calibration of a continuous temperature measurement system in a microwave cavity by infrared imaging. *Journal of Food Processing and Preservation*, 15: 331–337.

Jia, X. and Jolly, P. 1992. Simulation of microwave field and power distribution in a cavity by a three-dimensional finite element method. *Journal of Microwave Power and Electromagnetic Energy*, 27(1): 11–22.

Jones, P.L. 1986. High frequency dielectric heating in paper making. *Drying Technology*, 4(2): 217–244.

Kanellopoulos, V.N., Pichon, L., Razek, A., and Sekkak, A. 1995. A thermal and electromagnetic analysis in biological objects using 3D finite elements and absorbing boundary conditions. *IEEE Transactions on Magnetics*, 31(3): 1865–1868.

Khalil, H.K. 2002. *Nonlinear Systems*, 3rd ed. Prentice Hall, Upper Saddle River, NJ.

LeBail, A., Koutchma, T., and Ramaswamy, H.S. 2000. Modeling of temperature profiles under continuous tube-flow microwave and steam heating conditions. *Journal of Food Process Engineering*, 23: 1–24.

Lee, C.E. and Taylor, H.F. 1991. Fiber-optic Fabry–Perot temperature sensor using a low-coherence light source. *Journal of Lightwave Technology*, 9(I): 129–134.

Lin, Y.E., Anantheswaran, R.C., and Puri, V.M. 1995. Finite element analysis of microwave heating of solid foods. *Journal of Food Engineering*, 25: 85–112.

Lyons, B.E., Samulski, T.V., and Britt, R.H. 1985. Temperature measurements in high thermal gradients. 1. The effects of conduction. *International Journal of Radiation Oncology, Biology, physics*, 11: 951–62.

Marlin, T.E. 2000. *Process Control: Designing Processes and Control Systems for Dynamic Performance*, 2nd ed. McGraw-Hill Publishing Co., Boston, MA.

McFarlane, I. 1995. *Automatic Control of Food Manufacturing Processes*, 2nd ed. Chap. & Hall, Glasgow, UK.

Metaxas, A.C. and Meredith, R.J. 1983. *Industrial Microwave Heating*, Peter Peregrinus Ltd., London, UK.

Mittal, G.S. 1996. Computerized control systems in the food industry. Marcel Dekker, Inc., New York.

Morari, M., Lee, J.H., Garcia, C.E., Prett, D.M. 2005. *Model Predictive Control*, Prentice Hall, Englewood Cliffs.

Mudgett, R.E. 1985. Modeling microwave heating characteristics. In: *Microwaves in the Food Processing Industry*, R.V. Decareau (ed.). Academic Press, New York, pp. 38–56.

Mullin, J., and Bows, J. 1993. Temperature measurement during microwave cooking. *Food Additives and Contaminants*, 10(6): 663–672.

Nachman, M. and Turgeon, G. 1984. Heating patterns in multi-layered material exposed to microwaves. IEEE Transactions Microwave Theory and Techniques 32: 547.

Ogunnaike, B.A., Ray, W.H. 1994. *Process Dynamics, Modeling, and Control*, Oxford University Press, New York.

Ohlsson, T. 1993. In-flow microwave heating of pumpable foods. In: *Developments in Food Engineering: Proceedings of the 6th International Congress on Engineering and Food*. pp. 322–324.

Ohlsson, T., and Bengtsson, N.E. 1975. Dielectric food data for microwave sterilization. *Journal of Microwave Power*, 10, 93.

Pichon, L., Razek, A., and Sekkak, A. 1994. 3D FEM magneto-thermal analysis in microwave ovens. *IEEE Transactions on Magnetics*, 30(5): 3347–3350.

Reid, A.D., Gertner, M.R., and Sherar, M.D. 2001. Temperature measurement artefacts of thermocouples and fluoroptic probes during laser irradiation at 810 nm. *Physics in Medicine and Biology*, 46: N149–N157.

Rosenthal, I. 1992. *Electromagnetic Radiation in Food Science*, Springer-Verlag, Berlin.

Roussy, G. and Pearce, J.A. 1995. *Foundations and Industrial Applications of Microwave and Radio Frequency Fields: Physical and Chemical Processes*, John Wiley & Sons Ltd., Chichester, UK

Sabliov, C.M., Sandeep, K.P., and Simunovic, J. 2004. High frequency electromagnetism coupled with conductive heat transfer. A method to predict temperature profiles in materials heated in a focused microwave oven. In: *Microwave and Radio Frequency Applications: Proceedings of the 4th World Congress on Microwave and Radio Frequency Applications*. pp. 469–476.

Schiffmann, R.F. 1997. Principles of industrial microwave and RF heating. In: Microwaves: Theory and application in materials processing IV. D.E., Clark, W.H., Sutton, D.A., Lewis (eds.). The American Ceramic Society, Westerville, OH.

Seborg, D. E., Edgar, T.F., and Mellichamp, D.A. 2003. *Process Dynamics and Control*, John Wiley & Sons: New York.

Sekkak, A., Pichon, L., and Razek, A. 1994. 3-D FEM magneto-thermal analysis in microwave ovens. *IEEE Transactions on Magnetics*, 30(5): 3347–3350.

Soriano, V., Devece, C., de los Reyes, E. 1998. A finite element and finite difference formulation for microwave heating laminar material. *Journal of Microwave Power and Electromagnetic Energy*, 33(2): 67–76.

Spiegel, R.J. 1984. A review of numerical models for predicting the energy deposition and resultant thermal response of humans exposed to electromagnetic fields. *IEEE Transactions on Microwave Theory and Techniques*, 32: 730.

Sucher, M., and Fox, J. 1963. *Handbook of Microwave Measurements*, Vols I–III. Polytechnic Press, Wiley-Interscience, New York.

Sun, M.H., Wickersheim, K.A., and Kim, J. 1989. Fiberoptic temperature sensors in the medical setting. *Proceedings of SPIE*, 1067: 15–21.

Taflove, A., and Brodwin, M. E. 1975. Computation of the electromagnetic fields and induced temperatures within a model of the microwave irradiated human eye. IEEE *Transactions on Microwave Theory and Techniques*, 23: 888.

Vilayannur, R.S., Puri, V.M., and Anantheswaran, R.C. 1998. Size and shape effect on nonuniformity of temperature and moisture distributions in microwave heated food materials. Part I. Simulation. *Journal of Food Process Engineering*, 21: 209–233.

von Hippel, A.R. 1954. *Dielectric Materials and Application*, The Technology Press of M.I.T., John Wiley & Sons, Inc., New York and Chapman & Hall, Ltd., London, UK.

White, J.R. 1973. Why materials heat. Transactions of IMPI 1, Paper No. 4. 40–65.

Yakovlev, V.V. Modeling as a tool to design and optimize microwave applicators, In: *Fundamentals of Dielectric Heating and Applicator Design*, Short Course, Modena, Italy, September 12, 2005, 81 pp.

# INDEX

Note: Italicized page locators indicate a photo/figure; tables are noted with a *t*.




*Thermal Processing of Foods: Control and Automation*. Edited by K.P. Sandeep